A Key to Whitehead's
Process and Reality

A Key to
Whitehead's
*Process
and
Reality*

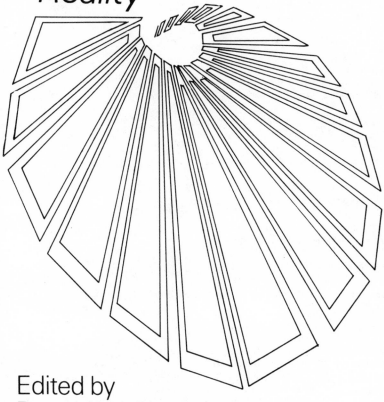

Edited by
Donald W. Sherburne

The University of Chicago Press
Chicago and London

This edition published by arrangement with The Free Press,
a Division of Macmillan Publishing Co., Inc.

The University of Chicago Press, Chicago 60637
The University of Chicago Press, Ltd., London

88 87 86 85 84 83   2 3 4 5

Library of Congress Cataloging in Publication Data

Sherburne, Donald W.
  A key to Whitehead's Process and reality.

  Includes index.
  1. Whitehead, Alfred North, 1861–1947.
Process and reality. 2. Cosmology. 3. Science—
Philosophy. 4. Process philosophy. 5. Organism
(Philosophy) I. Whitehead, Alfred North, 1861–1947.
Process and reality. II. Title.
B1674.W353P7638   1981   113   81-11661
ISBN: 0-226-75293-3 (pbk.)   AACR2

This book is dedicated to HENRY D. MINER of Leicester, Vermont. Farmer, Woodsman, Fisherman, Friend —He exemplifies to a high degree the qualities of ingenuity and sound common sense which, intuition tells me, would have led Whitehead (who vacationed and wrote in Vermont, among Vermonters) to approve of this dedication.

Acknowledgments

I thank both Vere C. Chappell and George L. Kline for numerous suggestions that improved the manuscript considerably, and I acknowledge, gratefully, a series of summer grants from Vanderbilt University which provided the freedom that allowed me to prepare this book.

Contents

LIST OF FIGURES

Introduction

PURPOSE AND DESIGN OF THIS BOOK

The interest in, and the influence of, the process philosophy of Alfred North Whitehead has mushroomed in the comparatively few years since his death in 1947,[1] and there is every indication that Whitehead's audience will continue to grow in succeeding years. But it is a disturbing fact that Whitehead is not to any great extent being taught to undergraduates in our colleges or being read by people in fields other than philosophy. The reason is not far to find: there has been a dearth of suitable material to introduce Whitehead to undergraduates and to the interested nonphilosopher. The present book aims to remedy this situation.

To understand the character of Whitehead's writings is both to understand the scarcity of introductory material and to grasp the particular sort of challenge that preparing this book has involved. The magnum opus, *central to an understanding of Whitehead's mature metaphysical position, is* Process and Reality *(commonly referred to as PR). PR is an extremely difficult book. One source of*

[1] The earlier commentaries by Dorothy Emmet and Ras-vihary Das have been superseded by a host of studies of Whitehead—e.g., those by Nathaniel Lawrence in 1956, Ivor Leclerc in 1958, William Christian in 1959, Wolfe Mays in 1959, Robert Palter in 1960, Donald Sherburne in 1961, and Victor Lowe in 1962. At the same time journals have carried numerous discussions of Whitehead and even more numerous analyses that presuppose the Whiteheadian metaphysics. Several vigorous symposiums and two volumes of commemorative essays marked the centennial of Whitehead's birth in 1961.

1

the difficulty is certainly the subtle character of the cosmology there developed. But further difficulty, the straw-that-breaks-the-reader's-back sort of difficulty, is occasioned by Whitehead's mode of exposition. As he states in his preface, "... the unity of treatment is to be looked for in the gradual development of the scheme, in meaning and in relevance, and not in the successive treatment of particular topics." This statement is a warning to the reader that he will not find a linear development in PR, *a beginning, a middle, and an end. Rather, he will encounter a weblike development that presupposes the whole system at the very beginning and recurs again and again to individual topics with the aim that "in each recurrence, these topics throw some new light on the scheme, or receive some new elucidation." The result is a book that in richness and suggestiveness is unsurpassed, but in opacity is monumental.*

But it is to PR *that one must go to encounter the essential Whitehead. It has been customary to send beginning students to White-head's* Science and the Modern World (SMW) *or his* Religion in the Making (RM). SMW *is organized in such a way that the first half of most chapters is a fascinating commentary on a given century of thought and the second half is a tight exposition of some aspect of Whitehead's own thought. Students all remember Whitehead's epigrammatic comparison of the Middle Ages with the eighteenth century: "The earlier period was the age of faith, based upon reason. In the later period, they let sleeping dogs lie: it was the age of reason, based upon faith. ... In comparing these epochs it is well to remember that reason can err, and that faith may be misplaced." But it is my experience that in reading* SMW *students learn very little about Whitehead's metaphysical system; his discussion of his own position is dense and cryptic and is fully intelligible only to one who has already mastered the scheme presented in* PR. *The same is true of* RM; *there the incredibly difficult third chapter, upon which all the rest depends, is absolutely incomprehensible to the student who has not previously acquainted himself with* PR. *Therefore,* SMW *and* RM *do not constitute an adequate introduction to Whitehead, but rather, depend upon a prior knowledge of* PR.

The problem this book faces is the problem of making the organic philosophy of PR *more accessible to students and nonphilosophers than it is in the original. The manner in which this has been done* can now be explained. Where Whitehead's development is weblike, the present book is linear; it does consist of "successive treatment of particular topics." The topics chosen seem to the editor to be those at the heart of Whitehead's system, and the ordering of these topics is designed to lead the student logically and coherently into the system.

This book, however, is not a series of comments about Whitehead. In it Whitehead himself speaks. In PR, Whitehead's comments on any given topic are scattered throughout dozens of passages. Here, Whitehead's scattered observations of PR are drawn together topic by topic to achieve maximum ease of comprehension while retaining the flavor and vigor of Whitehead's frequently brilliant prose.

This book does not give an exhaustive account of all aspects of Whitehead's philosophy, and it does not attempt critical evaluation of what it does present. But on the positive side, it does take the uninitiated into the heart of Whitehead's philosophy more quickly and easily than any other vehicle now in existence. This revisitation of Process and Reality will not of itself take the student into the storm of critical charge and countercharge currently raging about Whitehead's contribution to philosophy, but it will prepare the student both to understand such polemics and, more important, to plunge directly into the profound and rewarding experience of reading PR, SMW, and RM with insight and understanding.

Notes on the Text

Certain portions of this text present extended passages from Process and Reality, but much of the text consists of short passages from many different sections of PR woven into a sustained account of the successive aspects of Whitehead's system. To attempt to identify the source of each passage as it occurs in the text would break up the pages in a disconcerting manner. Consequently a section titled "Location in PR of Passages Quoted in this Book" appears in the back of the book; the reader can turn to this section at any time and quickly locate the source in PR of any passage in the text.

In practically all cases the sentence has been taken as an atomic

*unit not to be tampered with. Occasionally this rule has been vio-
lated. For example, the following appears in* PR: *"Perception is
frequently poorly analyzed. It is sometimes regarded as . . ." In the
present text this appears as follows: "Perception is sometimes regarded as . . ." In this case either brackets or an ellipsis would be
an unnecessary hindrance to smooth reading, and neither has been
used. Another instance of tampering within the sentence is this:
where Whitehead writes "For example, in one of the quotations
cited he writes: . . ." the present text substitutes "For example, he
writes: . . ." for the good reason that the cited quotation has not
appeared in this text. Where changes not of this simple, straightforward type have been made, brackets have been used.*

*Brackets in the text indicate insertions by the editor. Whitehead
uses few footnotes, and in most cases these have been inserted into
the text within parentheses. With only a few exceptions all footnotes are by the editor: exceptions are specifically attributed to
Whitehead.*

*A good many explanatory paragraphs by the editor appear in this
book. These paragraphs are italicized and clearly stand out from
Whitehead's text. Several diagrams not to be found in* PR *are included in these explanatory passages. Figures 2 and 3 are taken from
the editor's* A Whiteheadian Aesthetic *and are used with the per-
mission of the Yale University Press.*

*The Appendix, "In Defense of Speculative Philosophy," requires
a word of explanation. There are good reasons for making this unit
the first chapter of the book and good reasons for making it the last.
Its present status as an appendix is a compromise. The "Defense" is
useful if read at the beginning because it indicates the nature and
scope of the enterprise undertaken in* PR. *But it is more significant
if read at the end: the reader then has some feel for the concrete
character assumed by speculative philosophy at the hands of Whitehead, and he is therefore in a much better position to understand
and evaluate Whitehead's "Defense." The Appendix will probably
be of great help if read quickly at the beginning, but it should also
be read with care at the end, for it is only at the end that its self-
referential adequacy can be evaluated.*

*Whitehead's writing is liberally sprinkled with neologisms. It
takes time for the reader just becoming acquainted with these tech-*

nical terms to fix their meanings firmly in mind. To aid in this process a Glossary is appended to this book and the reader is urged to make use of it—it contains seventy items, cross-references, and examples and materials which frequently provide a substantial supplement to the main text.

The text of Process and Reality is in very poor condition. White-head, as I understand it, refused to have anything to do with the publishing process once he had completed a manuscript. He would have argued, I'm sure, that he didn't have time, that he had too many other books to write—and this would have been an irrefutable argument. But the result is that PR in particular is shot through and through with irritating typographical errors, apparent enough to the expert but bewildering to the neophyte, who finds it difficult to distinguish typos from neologisms. The editor, in collaboration with a group of Whitehead scholars, prepared a "Corrigenda for Process and Reality" which appeared as an Appendix to Alfred North Whitehead: Essays on His Philosophy (edited by George L. Kline and published by Prentice-Hall in 1963 as a Spectrum Book). This Corrigenda contained over two hundred items and more have since been identified. The present book has incorporated these corrections into its text. Whitehead's British spellings ("colour," "behaviour," "endeavour," "rôle," "analyse," and the like) and punctuation have been retained, although the editor uses Americanizations of these forms in his own discussions.

Chapter One

THE ACTUAL ENTITY

The concept of an actual entity is the central concept in White-head's system. This system is atomistic—i.e., like Democritus, White-head conceives of the world as composed of a vast number of microcosmic entities. But whereas Democritus is a materialist and views his atoms as inert bits of stuff, Whitehead presents an organic philosophy—each one of his atoms, termed "actual entities" or "actual occasions," is an organism that grows, matures, and perishes. The whole of Process and Reality (PR) is concerned with describing the characteristics of, and interrelationships between, actual entities.

The first three chapters will analyze actual entities in minute de-tail. When this analysis is completed the reader will know a great deal about how Whitehead describes the building blocks of the universe, but he will not yet have brought the system out of the realm of the microcosmic to confront his macrocosmic experience. Chapters Four through Seven do just that—i.e., they take the categories systematically presented in the first three chapters and put them into juxtaposition with ordinary experience, traditional philosophical problems, modern science, and religious intuitions. Chapters One through Three focus primarily upon the nature of the individual actual entity, which is a microcosmic entity; Chapters Four through Seven shift to the level of the macrocosmic, to an analysis of the aggregates of actual entities (termed "nexūs" and "societies") that are the objects of ordinary experience.

I

The Actual Entity

The positive doctrine of these lectures is concerned with the becoming, the being, and the relatedness of 'actual entities.' 'Actual entities'—also termed 'actual occasions'—are the final real things of which the world is made up. There is no going behind actual entities to find anything more real. They differ among themselves: God is an actual entity, and so is the most trivial puff of existence in far-off empty space.

The presumption that there is only one genus of actual entities constitutes an ideal of cosmological theory to which the philosophy of organism endeavours to conform. The description of the generic character of an actual entity should include God, as well as the lowliest actual occasion, though there is a specific difference between the nature of God and that of any occasion. But, though there are gradations of importance, and diversities of function, yet in the principles which actuality exemplifies all are on the same level. The final facts are, all alike, actual entities; and these actual entities are drops of experience, complex and interdependent.

[In his first *Meditation*], Descartes uses the phrase *res vera* in the same sense as that in which I have used the term 'actual.' It means 'existence' in the fullest sense of that term, beyond which there is no other. Descartes, indeed, would ascribe to God 'existence' in a generically different sense. In the philosophy of organism, as here developed, God's existence is not generically different from that of other actual entities, except that he is 'primordial' in a sense to be gradually explained.

'Concrescence' is the name for the process in which the universe of many things acquires an individual unity in a determinate relegation of each item of the 'many' to its subordination in the constitution of the novel 'one.' An actual occasion is nothing but the unity to be ascribed to a particular instance of concrescence. This concrescence is thus nothing else than the 'real internal constitution' of the actual occasion in question. The process itself is the constitution of the actual entity; in Locke's phrase, it is the 'real internal constitution' of the actual entity.

This is a theory of monads; but it differs from Leibniz's in that his monads change. In the organic theory, they merely *become*. Each monadic creature is a mode of the process of 'feeling' the world, of housing the world in one unit of complex feeling, in every way determinate. Such a unit is an 'actual occasion'; it is the ultimate creature derivative from the creative process.

Each actual entity is conceived as an act of experience arising out of data. The objectifications of other actual occasions form the given data from which an actual occasion originates. Each actual entity is a throb of experience including the actual world within its scope. It is a process of 'feeling' the many data, so as to absorb them into the unity of one individual 'satisfaction.' Here 'feeling' is the term used for the basic generic operation of passing from the objectivity of the data to the subjectivity of the actual entity in question. Feelings are variously specialized operations, effecting a transition into subjectivity. They replace the 'neutral stuff' of certain realistic philosophers. An actual entity is a process, and is not describable in terms of the morphology of a 'stuff.'

This word 'feeling' is a mere technical term; but it has been chosen to suggest that functioning through which the concrescent actuality appropriates the datum so as to make it its own. A feeling appropriates elements of the universe, which in themselves are other than the subject, and absorbs these elements into the real internal constitution of its subject by synthesizing them in the unity of an emotional pattern expressive of its own subjectivity. Feelings are 'vectors'; for they feel what is *there* and transform it into what is *here*. We thus say that an actual occasion is a concrescence effected by a process of feelings.

The philosophy of organism is a cell-theory of actuality. The cell is exhibited as appropriating, for the foundation of its own existence, the various elements of the universe out of which it arises. Each process of appropriation of a particular element is termed a prehension. I have adopted the term 'prehension' to express the activity whereby an actual entity effects its own concretion of other things. In Cartesian language, the essence of an actual entity consists solely in the fact that it is a prehending thing (i.e., a substance whose whole essence or nature is to prehend).

There are two species of prehensions, the 'positive species' and the 'negative species.' A 'feeling' belongs to the positive species of 'prehensions.' An actual entity has a perfectly definite bond with each item in the universe. This determinate bond is its prehension of that item. A negative prehension is the definite exclusion of that item from positive contribution to the subject's own real internal constitution. A positive prehension is the definite inclusion of that item into positive contribution to the subject's own real internal constitution. This positive inclusion is called its 'feeling' of that item. All actual entities in the actual world, relatively to a given actual entity as 'subject,' are necessarily 'felt' by that subject, though in general vaguely.

A feeling cannot be abstracted from the actual entity entertaining it. This actual entity is termed the 'subject' of the feeling. It is in virtue of its subject that the feeling is one thing. If we abstract the subject from the feeling we are left with many things. Thus a feeling is one aspect of its own subject.

II

Prehensions

Because an actual entity is constituted by its prehensions, these must now be considered in more detail. The analysis will center around two concepts, "datum" and "subjective form."

The first analysis of an actual entity, into its most concrete elements, discloses it to be a concrescence of prehensions, which have originated in its process of becoming. All further analysis is an analysis of prehensions. Every prehension consists of three factors: (a) the 'subject' which is prehending, namely, the actual entity in which that prehension is a concrete element; (b) the 'datum' which is prehended; (c) the 'subjective form' which is *how* that subject prehends that datum.

1. Datum

A 'simple physical feeling' entertained in one subject is a feeling for which the initial datum is another single actual entity, and the objective datum is another feeling entertained by the latter actual entity.

*Here a diagram will help the reader visualize the relationships
described. An actual entity has been seen to be constituted by its
feelings, or prehensions. Hence each actual entity will be portrayed
as a pie cut into pieces—the pie is the sum of its pieces as the actual
entity is the sum of its prehensions. In Figure 1, B is the concrescing,
subject actual entity, the entity in the process of becoming. A is an
actual entity in the immediate past of B, which is being prehended
by B. X is one of B's prehensions, the prehension that "reaches out"
to include A in B, the "vector" (from the Latin, vectus, past parti-
ciple of veho, to carry—used in mathematics to denote a line having
a fixed direction in space) that bears the A-ness of A into B. M, N,
and O are prehensions constitutive of A. N is the particular prehension in A selected by B to represent A, to objectify A, in B's concrescence. All the other prehensions in A are negatively prehended
by B; Y and Z represent negative prehensions that eliminate certain
aspects of A's constitution from relevance to B's feeling. Letters inserted in the following text correlate Whitehead's descriptive phrases
with the labels of Figure 1.*

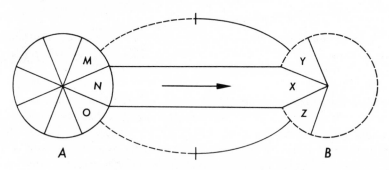

Figure 1. A Simple Physical Feeling.

Thus in a simple physical feeling [X] there are two actual entities
concerned. One of them [B] is the subject of that feeling, and the
other [A] is the *initial* datum of the feeling. A second feeling [N] is
also concerned, namely, the *objective* datum of the simple physical
feeling. This second feeling [N] is the 'objectification' of *its* subject
[A] for the subject [B] of the simple physical feeling [X]. The initial
datum [A] is objectified as being the subject of the feeling [N] which

is the objective datum: the objectification is the 'perspective' of the initial datum. The prehension [N] in one subject [A] becomes the objective datum for the prehension [X] in a later subject [B], thus objectifying the earlier subject [A] for the later subject [B]. Objectification relegates into irrelevance, or into a subordinate relevance, the full constitution of the objectified entity. Some real component in the objectified entity assumes the rôle of being how that particular entity is a datum in the experience of the subject.

A feeling [X] is the appropriation of some elements in the universe to be components in the real internal constitution of its subject [B]. The elements are the initial data; they are what the feeling feels. But they are felt under an abstraction. The process of the feeling involves negative prehensions [Y, Z] which effect elimination. There is a transition from the initial data to the objective datum effected by the elimination. The objective datum is the perspective of the initial data. Thus the initial data [A] are felt under a 'perspective' which is the objective datum [N] of the feeling [X].

In virtue of this elimination [Y, Z] the components of the objective datum [N] have become 'objects' intervening in the constitution of the subject [B] of the feeling [X]. In the phraseology of mathematical physics a feeling has a 'vector' character. A feeling [X] is the agency by which other things [A] are built into the constitution of its [X's] one subject in process of concrescence [B].

A simple physical feeling is an act of causation. The actual entity which is the initial datum [A] is the 'cause,' the simple physical feeling [X] is the 'effect,' and the subject entertaining the simple physical feeling [B] is the actual entity 'conditioned' by the effect. This 'conditioned' actual entity [B] will also be called the 'effect.' All complex causal action can be reduced to a complex of such primary components. Therefore simple physical feelings will also be called 'causal' feelings [or feelings of causal efficacy]. The 'power' of one actual entity on the other is simply how the former is objectified in the constitution of the other.

A simple physical feeling has the dual character of being the cause's feeling re-enacted for the effect as subject. By reason of this duplicity in a simple feeling there is a vector character which transfers the cause into the effect. It is a feeling *from* the cause which acquires the subjectivity of the new effect without loss of its original subjectivity in the cause. Simple physical feelings embody the reproductive character of nature, and also the objective immortality of the past. In virtue of these feelings time is the conformation of the immediate present to the past. Such feelings are conformal feelings.

2. *Subjective Form*

A feeling—i.e., a positive prehension—is essentially a transition effecting a concrescence. Its complex constitution is analysable into five factors which express what that transition consists of, and effects. The factors are: (i) the 'subject' which feels, (ii) the 'initial data' which are to be felt, (iii) the 'elimination' in virtue of negative prehensions, (iv) the 'objective datum' which is felt, (v) the 'subjective form' which is *how* that subject feels that objective datum. An actual entity, on its subjective side, is nothing else than what the universe is for it, including its own reactions. The reactions are the subjective forms of the feelings. There are many species of subjective forms, such as emotions, valuations, purposes, adversions, aversions, consciousness, etc.

The essential novelty of a feeling attaches to its subjective form. The initial data, and even the objective datum, may have served other feelings with other subjects. But the subjective form is the immediate novelty; it is how *that* subject is feeling that objective datum. There is no tearing this subjective form from the novelty of this concrescence. It is enveloped in the immediacy of its immediate present. The subjective form is the ingression of novel form peculiar to the new particular fact, and with its peculiar mode of fusion with the objective datum. In the becoming, it meets the 'data' which are selected from the actual world.

A feeling can be genetically described in terms of its process of origination, with its negative prehensions whereby its many initial data become its complex objective datum. In this process the subjective form originates, and carries into the feeling the way in which the feeling feels. The way in which the feeling feels expresses how the feeling came into being. It expresses the purpose which urged it forward, and the obstacles which it encountered, and the indeterminations which were dissolved by the originative decisions of the subject.

Physical feelings are always derived from some antecedent experient. Occasion B prehends occasion A as an antecedent subject experiencing a sensum with [a subjective form of] emotional intensity. B's subjective form of emotion is conformed to A's subjective form. Thus there is a vector transmission of emotional feeling of a sensum from A to B. In this way B feels the sensum as derived from A and feels it with an emotional form also derived from A. This is the most primitive form of the feeling of causal efficacy. In physics it is the transmission of a form of energy.

Apart from inhibitions or additions, weakenings or intensifications, due to the history of its production, the subjective form of a physical feeling [X] is re-enaction of the subjective form of the feeling felt [N]. The subjective form, amid its own original elements, always involves reproduction of the pattern of the objective datum. Thus the cause passes on its feeling to be reproduced by the new subject as its own, and yet as inseparable from the cause. There is a flow of feeling. But the re-enaction is not perfect. The feeling is always novel in reference to its data; since its subjective form, though it must always have reproductive reference to the data, is not wholly determined by them.

The cause [A] is objectively in the constitution of the effect [B], in virtue of being the feeler of the feeling [N] reproduced in the effect [as X] with partial equivalence of subjective form. The reason why the cause [A] is [merely] objectively in the effect, is that the cause's feeling [N] cannot, as a feeling, be abstracted from its subject [A] which is the cause.

III

Satisfaction, Superject, and Objective Immortality

A cluster of three notions—"satisfaction," "superject," and "objective immortality"—elucidates the completion of the process constitutive of the actual entity. An actual entity initiates its process by prehending many other datum occasions in its causal past. Then occur the complexities of concrescence (to be analyzed in detail in Chapter Three) during which the many initial single feelings of the world are integrated into one complex feeling of that occasion's

*actual world. This final, complex, integrated feeling is the satisfac-
tion of that occasion. The satisfaction closes up, completes,
concludes the actual entity. The satisfaction of an actual entity determines its character as superject—i.e., the character it has as objectively immortal, its character as an object encountered as initial
datum by succeeding actual entities. The discussion here of the
notion of "satisfaction" is a preliminary discussion; the topic receives fuller treatment in Section VI of Chapter Three.*

An actual entity is a process in the course of which many operations with incomplete subjective unity terminate in a completed
unity of operation, termed the 'satisfaction.' The actual entity
terminates its becoming in one complex feeling involving a completely determinate bond with every item in the universe, the bond
being either a positive or a negative prehension. This termination is
the 'satisfaction' of the actual entity.

The process of the concrescence is a progressive integration of
feelings controlled by their subjective forms. In this synthesis, feelings of an earlier phase sink into the components of some more complex feeling of a later phase. Thus each phase adds its element of
novelty, until the final phase in which the one complex 'satisfaction'
is reached. This process of the integration of feeling proceeds until
the concrete unity of feeling is obtained. In this concrete unity all
indetermination as to the realization of possibilities has been eliminated. The many entities of the universe, including those originating
in the concrescence itself, find their respective rôles in this final
unity. The 'satisfaction' is the culmination of the concrescence into
a completely determinate matter of fact.

The attainment of a peculiar definiteness is the final cause which
animates a particular process; and its attainment halts its process,
so that by transcendence it passes into its objective immortality as a
new objective condition added to the riches of definiteness attainable. It enjoys an objective immortality in the future beyond itself.

The peculiarity of an actual entity is that it can be considered
both 'objectively' and 'formally.' The 'formal' aspect is functional
so far as that actual entity is concerned: by this it is meant that the
process involved is immanent in it. The 'formal' reality of the actuality in question belongs to its process of concrescence and not to its

'satisfaction.' The 'objective' aspect is morphological so far as that actual entity is concerned: by this it is meant that the process involved is transcendent relatively to it, so that the *esse* of its satisfaction is *sentiri*. The objective consideration is pragmatic. It is the consideration of the actual entity in respect to its consequences.

The terminal unity of operation, here called the 'satisfaction,' embodies what the actual entity is beyond itself. In Locke's phraseology, the 'powers' of the actual entity are discovered in the analysis of the satisfaction. It is the actual entity as a definite, determinate, settled fact, stubborn and with unavoidable consequences. Its own process, which is its own internal existence, has evaporated, worn out and satisfied; but its effects are all to be described in terms of its 'satisfaction.' The 'effects' of an actual entity are its interventions in concrescent processes other than its own. Any entity, thus intervening in processes transcending itself, is said to be functioning as an 'object.' It is the one general metaphysical character of all entities of all sorts, that they function as objects. It is this metaphysical character which constitutes the solidarity of the universe.

An actual entity is to be conceived both as a subject presiding over its own immediacy of becoming, and a superject which is the atomic creature exercising its function of objective immortality. It has become a 'being'; and it belongs to the nature of every 'being' that it is a potential for every 'becoming.' To be actual must mean that all actual things are alike objects, enjoying objective immortality in fashioning creative actions; and that all actual things are subjects, each prehending the universe from which it arises.

It is fundamental to the metaphysical doctrine of the philosophy of organism, that the notion of an actual entity as the unchanging subject of change is completely abandoned. An actual entity is at once the subject experiencing and the superject of its experiences. It is subject-superject, and neither half of this description can for a moment be lost sight of. The term 'subject' will be mostly employed when the actual entity is considered in respect to its own real internal constitution. But 'subject' is always to be construed as an abbreviation of 'subject-superject.'

The term 'subject' has been retained because in this sense it is familiar in philosophy. But it is misleading. The philosophies of substance presuppose a subject which then encounters a datum, and then reacts to the datum. The philosophy of organism presupposes a datum which is met with feelings, and progressively attains the unity of a subject. But with this doctrine, 'superject' would be a better term than 'subject.' The subject-superject is the purpose of the process originating the feelings. The feelings are inseparable from the end at which they aim; and this end is the feeler. The feelings aim at the feeler, as their final cause. The feelings are what they are in order that their subject may be what it is. Then transcendently, since the subject is what it is in virtue of its feelings, it is only by means of its feelings that the subject objectively conditions the creativity transcendent beyond itself. In our own relatively high grade of human existence, this doctrine of feelings and their subject is best illustrated by our notion of moral responsibility. The subject is responsible for being what it is in virtue of its feelings. It is also derivatively responsible for the consequences of its existence because they flow from its feelings.

If the subject-predicate form of statement be taken to be metaphysically ultimate, it is then impossible to express this doctrine of feelings and their superject. It is better to say that the feelings *aim at* their subject, than to say that they *are aimed at* their subject. For the latter mode of expression removes the subject from the scope of the feeling and assigns it to an external agency. Thus the feeling would be wrongly abstracted from its own final cause. This final cause is an inherent element in the feeling, constituting the unity of that feeling. An actual entity feels as it does feel in order to be the actual entity which it is. In this way an actual entity satisfies Spinoza's notion of substance: it is *causa sui*.

An 'actual entity' is a *res vera* in the Cartesian sense of that term; it is a Cartesian 'substance,' and not an Aristotelian 'primary substance.' But Descartes retained in his metaphysical doctrine the Aristotelian dominance of the category of 'quality' over that of 'relatedness.' In these lectures 'relatedness' is dominant over 'quality.' All relatedness has its foundation in the relatedness of actualities; and such relatedness is wholly concerned with the appropriation of the dead by the living—that is to say, with 'objective immortality' whereby what is divested of its own living immediacy becomes a real component in other living immediacies of becoming. This is the doctrine that the creative advance of the world is the becoming, the perishing, and the objective immortalities of those things which jointly constitute *stubborn fact* [i.e., actual entities].

This doctrine of organism is the attempt to describe the world as a process of generation of individual actual entities, each with its own absolute self-attainment. This concrete finality of the individual is nothing else than a decision referent beyond itself. The 'perpetual perishing' of individual absoluteness is thus foredoomed. But the 'perishing' of absoluteness is the attainment of 'objective immortality.'

The notion of 'satisfaction' is the notion of the 'entity as concrete' abstracted from the 'process of concrescence'; it is the outcome separated from the process, thereby losing the actuality of the atomic entity, which is both process and outcome. 'Satisfaction' provides the individual element in the composition of the actual entity—that element which has led to the definition of substance as 'requiring nothing but itself in order to exist.' But the 'satisfaction' is the 'superject' rather than the 'substance' or the 'subject.' It closes up the entity; and yet is the superject adding its character to the creativity whereby there is a becoming of entities superseding the one in question. This satisfaction is the attainment of something individual to the entity in question. It cannot be construed as a component contributing to its own concrescence; it is the ultimate fact, individual to the entity.

IV

The Ontological Principle

The importance of the concept of an actual entity is emphasized by what Whitehead terms "the ontological principle."

Every condition to which the process of becoming conforms in any particular instance, has its reason *either* in the character of some actual entity in the actual world of that concrescence, *or* in the character of the subject which is in process of concrescence. This is the 'ontological principle.' This ontological principle means that actual entities are the only *reasons;* so that to search for a *reason* is to search for one or more actual entities.

The 'ontological principle' broadens and extends a general principle laid down by John Locke in his *Essay* (Bk. II, Ch. XXIII, Sect. 7), when he asserts that 'power' is 'a great part of our complex ideas of substances.' The notion of 'substance' is transformed into that of 'actual entity'; and the notion of 'power' is transformed into the principle that the reasons for things are always to be found in the composite nature of definite actual entities—in the nature of God for reasons of the highest absoluteness, and in the nature of definite temporal actual entities for reasons which refer to a particular environment. The ontological principle can be summarized as: no actual entity, then no reason.

The actual world is built up of actual occasions; and by the ontological principle whatever things there are in any sense of 'existence,' are derived by abstraction from actual occasions. Apart from the experiences of subjects there is nothing, nothing, nothing, bare nothingness. The most general term 'thing'—or, equivalently, 'entity'—means nothing else than to be one of the 'many' which find their niches in each instance of concrescence. Each instance of concrescence *is itself* the novel individual 'thing' in question. There are not 'the concrescence' *and* the 'novel thing': when we analyse the novel thing we find nothing but the concrescence. 'Actuality' means nothing else than this ultimate entry into the concrete, in abstraction from which there is mere nonentity. In other words, abstraction from the notion of 'entry into the concrete' is a self-contradictory notion, since it asks us to conceive a thing as not a thing.

Descartes does not explicitly frame the definition of actuality in terms of the ontological principle, that actual occasions form the ground from which all other types of existence are derivative and abstracted; but he practically formulates an equivalent in subject-predicate phraseology, when he writes: "For this reason, when we perceive any attribute, we therefore conclude that some existing thing or substance to which it may be attributed, is necessarily present" (*Principles of Philosophy,* Part I, 52). For Descartes the word 'substance' is the equivalent of my phrase 'actual occasion.' I refrain from the term 'substance,' for one reason because it suggests the subject-predicate notion; and for another reason because Descartes and Locke permit their substances to undergo adventures of changing qualifications, and thereby create difficulties.

For rationalistic thought, the notion of 'givenness' carries with it a reference beyond the mere data in question. It refers to a 'decision' whereby what is 'given' is separated off from what for that occasion is 'not given.' This element of 'givenness' in things implies some activity procuring limitation. The word 'decision' does not here imply conscious judgment, though in some 'decisions' consciousness will be a factor. The word is used in its root sense of a 'cutting off.' The ontological principle declares that every decision is referable to one or more actual entities, because in separation from actual entities there is nothing, merely nonentity—'The rest is silence.'

The ontological principle asserts the relativity of decision; whereby every decision expresses the relation of the actual thing, *for which* a decision is made, to an actual thing *by which* that decision is made. But 'decision' cannot be construed as a casual adjunct of an actual entity. It constitutes the very meaning of actuality. An actual entity arises from decisions *for* it, and by its very existence provides decisions *for* other actual entities which supersede it. Thus the ontological principle is the first stage in constituting a theory embracing the notions of 'actual entity,' 'givenness,' and 'process.' Just as 'potentiality for process' is the meaning of the more general term 'entity,' or 'thing'; so 'decision' is the additional meaning imported by the word 'actual' into the phrase 'actual entity.' 'Actuality' is the decision amid 'potentiality.' It represents stubborn fact which cannot be evaded. The real internal constitution of an actual entity progressively constitutes a decision conditioning the creativity which transcends that actuality.

Chapter Two

THE FORMATIVE ELEMENTS

I
Eternal Objects

In the last few sentences of Chapter One reference was made by Whitehead to "potentiality" and to "creativity." These terms refer to two of the three formative elements, namely, eternal objects and creativity. The third formative element is God. The status of each of these formative elements, and the relationships holding among them, will be clarified gradually. It is from their mutual interaction that the universe of actual entities emerges—hence the appropriateness of Whitehead's designation, "formative" elements. To understand the functioning of the formative elements is to understand more clearly the nature of the actual entities they "form."

The first formative element to be considered is the element of pure potentiality in the universe, the eternal objects.

By stating my belief that the train of thought in these lectures is Platonic, I mean that if we had to render Plato's general point of view with the least changes made necessary by the intervening two thousand years of human experience in social organization, in aesthetic attainments, in science, and in religion, we should have to set about the construction of a philosophy of organism. In such a philosophy the actualities constituting the process of the world are conceived as exemplifying the ingression (or 'participation') of other things which constitute the potentialities of definiteness for any actual existence. The things which are temporal [actual entities] arise by their participation in the things which are eternal [eternal objects].

But these lectures are not an exegesis of Plato's writings; the entities in question are not necessarily restricted to those which he would recognize as 'forms.' Also the term 'idea' has a subjective suggestion in modern philosophy, which is very misleading for my present purposes; and in any case it has been used in many senses and has become ambiguous. The term 'essence,' as used by the Critical Realists, also suggests their use of it, which diverges from what I intend. Accordingly, by way of employing a term devoid of misleading suggestions, I use the phrase 'eternal object.' If the term 'eternal objects' is disliked, the term 'potentials' would be suitable. The eternal objects are the pure potentials of the universe; and the actual entities differ from each other in their realization of potentials. Any entity whose conceptual recognition does not involve a necessary reference to any definite actual entities of the temporal world is called an 'eternal object.'

An eternal object is always a potentiality for actual entities; but in itself, as conceptually felt, it is neutral as to the fact of its physical ingression in any particular actual entity of the temporal world. 'Potentiality' is the correlative of 'givenness.' The meaning of 'givenness' is that what *is* 'given' might not have been 'given'; and that what *is not* 'given' *might have been* 'given.' It is evident that 'givenness' and 'potentiality' are both meaningless apart from a multiplicity of potential entities. These potentialities are the 'eternal objects.' Apart from 'potentiality' and 'givenness,' there can be no nexus of actual things in process of supersession by novel actual things. The alternative is a static monistic universe, without unrealized potentialities; since 'potentiality' is then a meaningless term.

The functioning of an eternal object in the self-creation of an actual entity is the 'ingression' of the eternal object in the actual entity. An eternal object can be described only in terms of its potentiality for 'ingression' into the becoming of actual entities;

and its analysis only discloses other eternal objects. It is a pure potential. The term 'ingression' refers to the particular mode in which the potentiality of an eternal object is realized in a particular actual entity, contributing to the definiteness of that actual entity.

Actual occasions in their 'formal' constitutions are devoid of all indetermination. Potentiality has passed into realization. They are complete and determinate matter of fact, devoid of all indecision. But eternal objects involve in their own natures indecision. They are, like all entities, potentials for the process of becoming. Their ingression expresses the *definiteness* of the actuality in question. But their own natures do not in themselves disclose in what actual entities this potentiality of ingression is realized.

An eternal object in abstraction from any one particular actual entity is a potentiality for ingression into actual entities. In its ingression into any one actual entity, either as relevant or as irrelevant, it retains its potentiality of indefinite diversity of modes of ingression, a potential indetermination rendered determinate in this instance. The definite ingression into a particular actual entity is not to be conceived as the sheer evocation of that eternal object from 'not-being' into 'being'; it is the evocation of determination out of indetermination. Potentiality becomes reality; and yet retains its message of alternatives which the actual entity has avoided. In the constitution of an actual entity:—whatever component is red, might have been green; and whatever component is loved, might have been coldly esteemed.

Prehensions of actual entities—i.e., prehensions whose data involve actual entities—are termed 'physical prehensions'; and prehensions of eternal objects are termed 'conceptual prehensions' [or 'conceptual feelings']. In the technical phraseology of these lectures, a conceptual feeling is a feeling whose 'datum' is an eternal object. Analogously a negative prehension is termed 'conceptual,' when its datum is an eternal object.

A conceptual feeling is feeling an eternal object in the primary metaphysical character of being an 'object,' that is to say, feeling its *capacity* for being a realized determinant of process. Immanence and transcendence are the characteristics of an object: as a realized determinant it is immanent; as a capacity for determination it is transcendent; in both rôles it is relevant to something not itself.

There is no character belonging to the actual apart from its exclusive determination by selected eternal objects. The definiteness of the actual arises from the exclusiveness of eternal objects in their function as determinants. If the actual entity be *this,* then by the nature of the case it is not *that* or *that.* The fact of incompatible alternatives is the ultimate fact in virtue of which there is definite character. A conceptual feeling is the feeling of an eternal object in respect to its general capacity as a determinant of character, including thereby its capacity of exclusiveness.

An actual entity in the actual world of a subject *must* enter into the concrescence of that subject by *some* simple causal feeling, however vague, trivial, and submerged. Negative prehensions may eliminate its distinctive importance. But in some way, by some trace of causal feeling, the remote actual entity is prehended positively. In the case of an eternal object, there is no such necessity. In any given concrescence, it may be included positively by means of a conceptual feeling; but it may be excluded by a negative prehension. The actualities *have* to be felt, while the pure potentials *can* be dismissed. So far as concerns their functionings as objects, this is the great distinction between an actual entity and an eternal object. The one is stubborn matter of fact; and the other never loses its 'accent' of potentiality.

This initial account of the doctrine of eternal objects can now be used to make a very important point. Chapter One made it clear that every concrescing actual entity involves other actual entities in its process of becoming—i.e., other actual entities serve as data for concrescing subjects. A prehension is a vector in that it bears along what is there, transforming it into what is here. Eternal objects play a crucial role in effecting this transformation. To understand this role is to understand a Whiteheadian principle as basic as the "ontological principle"—namely, the "principle of relativity."

The potentiality for being an element in a real concrescence of many entities into one actuality, is the one general metaphysical character attaching to all entities, actual and non-actual [i.e., to actual entities and eternal objects]. Every item in its universe is involved in each concrescence. In other words, it belongs to the nature of a 'being' that it is a potential for every 'becoming.' This is the 'principle of relativity.' It asserts that the notion of an 'entity' means 'an element contributory to the process of becoming.'

The principle of universal relativity directly traverses Aristotle's dictum, '(A substance) is not present in a subject.' On the contrary, according to this principle an actual entity *is* present in other actual entities. In fact if we allow for degrees of relevance, and for negligible relevance, we must say that every actual entity is present in every other actual entity. The philosophy of organism is mainly devoted to the task of making clear the notion of 'being present in another entity.' This phrase is here borrowed from Aristotle: it is not a fortunate phrase, and in subsequent discussion it will be replaced by the term 'objectification.' The functioning of one actual entity in the self-creation of another actual entity is the 'objectification' of the former for the latter actual entity. The Aristotelian phrase suggests the crude notion that one actual entity is added to another *simpliciter*. This is not what is meant. One rôle of the eternal objects is that they are those elements which express how any one actual entity is constituted by its synthesis of other actual entities.

The organic philosophy does not hold that the 'particular existents' [i.e., actual entities] are prehended apart from universals [i.e., eternal objects]; on the contrary, it holds that they are prehended by the mediation of universals. In other words, each actuality is prehended by means of some element of its own definiteness. Eternal objects determine *how* the world of actual entities enters into the constitution of each one of its members via its feelings. The eternal objects function by introducing the multiplicity of actual entities as constitutive of the actual entity in question. For the philosophy of organism, the primary data are always actual entities absorbed into feeling in virtue of certain universals shared alike by the objectified actuality and the experient subject.

A simple physical feeling enjoys a characteristic which has been variously described as 're-enaction,' 'reproduction,' and 'conformation.' This characteristic can be more accurately explained in terms of the eternal objects involved. There are eternal objects determinant of the definiteness of the objective datum which is the 'cause,' and eternal objects determinant of the definiteness of the subjective form belonging to the 'effect.' When there is re-enaction there is one eternal object with two-way functioning, namely, as partial determinant of the objective datum, and as partial determinant of the subjective form. In this two-way rôle, the eternal object is functioning relationally between the initial data on the one hand and the concrescent subject on the other. In the conformal feelings the *how* of feeling reproduces what is felt. Some conformation is necessary as a basis of vector transition, whereby the past is synthesized with the present. The one eternal object in its two-way function, as a determinant of the datum and as a determinant of the subjective form, is thus relational. In this sense the solidarity of the universe is based on the relational functioning of eternal objects.

II
God

The second formative element to be considered is God. The simplest way to introduce the Whiteheadian concept of God is to apply the ontological principle to the notion of a realm of eternal objects. The result reveals much about the second formative element, and also reveals its relationship to the first formative element.

The scope of the ontological principle is not exhausted by the corollary that 'decision' must be referable to an actual entity. Everything must be somewhere; and here 'somewhere' means 'some actual entity.' It is a contradiction in terms to assume that some explanatory fact can float into the actual world out of nonentity. Nonentity is nothingness. Every explanatory fact refers to the decision and to the efficacy of an actual thing. Accordingly the general potentiality of the universe must be somewhere; since it retains its proximate relevance to actual entities for which it is unrealized. This 'somewhere' is the non-temporal actual entity.

The things which are temporal [actual occasions] arise by their participation in the things which are eternal [eternal objects]. The two sets are mediated by a thing which combines the actuality of what is temporal with the timelessness of what is potential. This final entity is the divine element in the world, by which the barren inefficient disjunction of abstract potentialities obtains primordially the efficient conjunction of ideal realization. By this recognition of the divine element the general Aristotelian principle is maintained that, apart from things that are actual, there is nothing—nothing either in fact or in efficacy.

The endeavor to understand eternal objects in complete abstraction from the actual world results in reducing them to mere undifferentiated nonentities. Accordingly the differentiated relevance of eternal objects to each instance of the creative process requires their conceptual realization in the primordial nature of God. The general relationships of eternal objects to each other, relationships of diversity and of pattern, are their relationships in God's conceptual realization. Apart from this realization, there is mere isolation indistinguishable from nonentity.

In what sense can unrealized abstract form be relevant? What is its basis of relevance? 'Relevance' must express some real fact of togetherness among forms. The ontological principle can be expressed as: All real togetherness is togetherness in the formal constitution of an actuality. So if there be a relevance of what in the temporal world is unrealized, the relevance must express a fact of togetherness in the formal constitution of a non-temporal actuality. But by the principle of relativity there can only be one non-derivative actuality, unbounded by its prehensions of an actual world. Unfettered conceptual valuation, 'infinite' in Spinoza's sense of that term, is only possible once in the universe; since that creative act is objectively immortal as an inescapable condition characterizing creative action. Such a primordial superject of creativity achieves, in its unity of satisfaction, the complete conceptual valuation of all eternal objects. This is the ultimate, basic adjustment of the togetherness of eternal objects on which creative order depends. It is the conceptual adjustment of all appetites in the form of aversions and adversions. It constitutes the meaning of relevance. Its status as an actual efficient fact is recognized by terming it the 'primordial nature of God.'

The primordial created fact is the unconditioned conceptual valuation of the entire multiplicity of eternal objects. This is the 'primordial nature' of God. He is the unconditioned actuality of conceptual feeling at the base of things; so that, by reason of this primordial actuality, there is an order in the relevance of eternal objects to the process of creation. He is the actual entity in virtue of which the *entire* multiplicity of eternal objects obtains its graded relevance to each stage of concrescence. Apart from God, there could be no relevant novelty. His unity of conceptual operations is a free creative act, untrammelled by reference to any particular course of things. It is deflected neither by love, nor by hatred, for what in fact comes to pass. The *particularities* of the actual world presuppose *it;* while *it* merely presupposes the *general* metaphysical character of creative advance, of which it is the primordial exemplification. The primordial nature of God is the acquirement by creativity of a primordial character.

This ideal realization of potentialities in a primordial actual entity constitutes the metaphysical stability whereby the actual process exemplifies general principles of metaphysics, and attains the ends proper to specific types of emergent order. By reason of the actuality of this primordial valuation of pure potentials, each eternal object has a definite, effective relevance to each concrescent process. Apart from such orderings, there would be a complete disjunction of eternal objects unrealized in the temporal world. Novelty would be meaningless, and inconceivable. By reason of this complete valuation, the objectification of God in each derivate actual entity results in a graduation of the relevance of eternal objects to the concrescent phases of that derivate occasion. There will be additional ground of relevance for select eternal objects by reason of their ingression into derivate actual entities belonging to the actual world of the concrescent occasion in question. But whether or no this be the case, there is always the definite relevance derived from God. Thus possibility which transcends realized temporal matter of fact has a real relevance to the creative advance. Apart from God, eternal objects unrealized in the actual world would be relatively non-existent for the concrescence in question. For effective relevance requires agency of comparison, and agency belongs exclusively to actual occasions. This divine ordering is itself matter of fact, thereby conditioning creativity. It is here termed 'God'; because the contemplation of our natures, as enjoying real feelings derived from the timeless source of all order, acquires that 'subjective form' of refreshment and companionship at which religions aim.

1.   God and Subjective Aim

The purpose of this section on God is to understand his function-ing as a formative element. And now that certain general remarks about God have been made, his role as formative element can be indicated more precisely. To understand God as formative element is to understand the part played by his primordial nature in the formation of the subjective aim of each and every actual occasion in the temporal world. Subjective aim must not be confused with subjective form. Subjective form is how a feeling is felt by the con-crescing subject of that feeling. Subjective aim concerns the direc-tion to be taken by the concrescing subject in the process that consti-tutes the very being of that subject. The subject does not exist prior to its concrescence, it comes into being with its concrescence, it is its concrescence—its being is its becoming. Every concrescence, which is causa sui, faces the question of what sort of entity it will make itself. The subjective aim, derived from God, is a lure (to be more or less completely followed) toward that way of becoming which is most in line with God's own aim of creating intensity of harmonious feeling in the world. The items in the Glossary for Subjective aim and God illuminate these relationships.

This doctrine of the inherence of the subject in the process of its production requires that in the primary phase of the subjective process there be a conceptual feeling of subjective aim. The immedi-acy of the concrescent subject is constituted by its living aim at its own self-constitution. The initial stage of its aim is an endow-ment which the subject inherits from the inevitable ordering of things, conceptually realized in the nature of God. Each temporal entity derives from God its basic conceptual aim, relevant to its actual world, yet with indeterminations awaiting its own decisions. Thus the initial stage of the aim is rooted in the nature of God, and its completion depends on the self-causation of the subject-superject.

God is the principle of concretion; namely, he is that actual entity from which each temporal concrescence receives that initial aim from which its self-causation starts. That aim determines the initial gradations of relevance of eternal objects for conceptual feeling; and constitutes the autonomous subject in its primary phase of feelings with its initial conceptual valuations, and with its initial physical purposes. Thus the transition of the creativity from an actual world to the correlate novel concrescence is conditioned by the relevance of God's all-embracing conceptual valuations to the particular possibilities of transmission from the actual world, and by its relevance to the various possibilities of initial subjective form available for the initial feelings. If we prefer the phraseology, we can say that God and the actual world jointly constitute the character of the creativity for the initial phase of the novel concrescence. The subject, thus constituted, is the autonomous master of its own concrescence into subject-superject. It passes from a subjective aim in concrescence into a superject with objective immortality. At any stage it is subject-superject.

2. Coherence of the Concept "God"

It might appear that the concept of God is an ad hoc *creation that, although it serves to link actuality and potentiality, is not itself correlated with other basic principles of the system. Such unrelatedness of basic principles results in incoherence. Whitehead is aware that the charge of incoherence might be brought against him, and in the following passages he argues that he has not introduced a mere* deus ex machina *unrelated to the other elements of his philosophy. The second and third aspects of the threefold character of God introduced in these passages, namely, his consequent nature and his superjective nature, may not be clearly understood by the reader. This is of no concern at the moment for they will be returned to and explicated in detail in Chapter Seven. It is enough at present to sense that they constitute evidence for Whitehead's claim that God is neither unrelated to nor an exception to the principles of the system.*

God is not to be treated as an exception to all metaphysical principles, invoked to save their collapse. He is their chief exemplification. The presumption that there is only one genus of actual entities constitutes an ideal of cosmological theory to which the philosophy of organism endeavours to conform. The description of the generic character of an actual entity should include God, as well as the lowliest actual occasion, though there is a specific difference between the nature of God and that of any occasion.

An actual entity has a threefold character. (i) It has the character 'given' for it by the past; the 'objectifications' of the actual entities in the actual world, relative to a definite actual entity, constitute the efficient causes out of which *that* actual entity arises. (ii) It has the subjective character aimed at in its process of concrescence; the 'subjective aim' at 'satisfaction' constitutes the final cause, or lure, whereby there is determinate concrescence. (iii) It has the superjective character, which is the pragmatic value of its specific satisfaction qualifying the transcendent creativity; that attained 'satisfaction' remains as an element in the content of creative purpose.

In the case of the primordial actual entity, which is God, there is no past. Thus the ideal realization of conceptual feeling takes the precedence. There is still, however, the same threefold character: (i) The 'primordial nature' of God is the concrescence of an unity of conceptual feelings, including among their data all eternal objects. The concrescence is directed by the subjective aim, that the subjective forms of the feelings shall be such as to constitute the eternal objects into relevant lures of feeling severally appropriate for all realizable basic conditions. (ii) The 'consequent nature' of God is the physical prehension by God of the actualities of the evolving universe. His primordial nature directs such perspectives of objectification so that each novel actuality in the temporal world contributes such elements as it can to a realization in God free from inhibitions of intensity by reason of discordance. (iii) The 'superjective' nature of God is the character of the pragmatic value of his specific satisfaction qualifying the transcendent creativity in the various temporal instances.

It is to be noted that every actual entity, including God, is something individual for its own sake; and thereby transcends the rest of actuality. And also it is to be noted that every actual entity, including God, is a creature transcended by the creativity which it qualifies. A temporal occasion in respect to the second element of its character, and God in respect to the first element of his character satisfy Spinoza's definition of substance, that it is *causa sui*. To be *causa sui* means that the process of concrescence is its own reason for the decision in respect to the qualitative clothing of feelings. It is finally responsible for the decision by which any lure for feeling is admitted to efficiency. The freedom inherent in the universe is constituted by this element of self-causation.

3. Summary and Transition to Creativity

God is the organ of novelty, aiming at intensification. He is the lure for feeling, the eternal urge of desire. The primary element in the 'lure for feeling' is the subject's prehension of the primordial nature of God. His particular relevance to each creative act as it arises from its own conditioned standpoint in the world, constitutes him the initial 'object of desire' establishing the initial phase of each subjective aim. Apart from the intervention of God, there could be nothing new in the world, and no order in the world. The course of creation would be a dead level of ineffectiveness, with all balance and intensity progressively excluded by the cross currents of incompatibility. The novel feelings derived from God are the foundations of progress.

This is the conception of God, according to which he is considered as the outcome of creativity, as the foundation of order, and as the goad towards novelty. 'Order' and 'novelty' are but the instruments of his subjective aim which is the intensification of 'formal immediacy.' Thus God's purpose in the creative advance is the evocation of intensities. This function of God is analogous to the remorseless working of things in Greek and in Buddhist thought. The initial aim is the best for that *impasse*. But if the best be bad, then the ruthlessness of God can be personified as *Atè*, the goddess of mischief. The chaff is burnt. What is inexorable in God, is valuation as an aim towards 'order'; and 'order' means 'society permissive of actualities with patterned intensity of feeling arising from adjusted contrasts.'

God can be termed the creator of each temporal actual entity. But the phrase is apt to be misleading by its suggestion that the ultimate creativity of the universe is to be ascribed to God's volition. The true metaphysical position is that God is the aboriginal instance of this creativity, and is therefore the aboriginal condition which qualifies its action. Viewed as primordial, he is the unlimited conceptual realization of the absolute wealth of potentiality. In this aspect, he is not *before* all creation, but *with* all creation. It is the function of actuality to characterize the creativity, and God is the eternal primordial character. But of course, there is no meaning to 'creativity' apart from its 'creatures,' and no meaning to 'God' apart from the creativity and the 'temporal creatures,' and no meaning to the temporal creatures apart from 'creativity' and 'God.'

III

Creativity

The third formative element is creativity. The concluding paragraph of the preceding section cryptically adumbrates the relationship between God and creativity. A more careful account of this elusive but crucial concept, creativity, is now required.

In all philosophic theory there is an ultimate which is actual in virtue of its accidents. It is only then capable of characterization through its accidental embodiments, and apart from these accidents is devoid of actuality. In the philosophy of organism this ultimate is termed 'creativity'; and God is its primordial, nontemporal accident. The creativity is not an external agency with its own ulterior purposes. In monistic philosophies, Spinoza's or absolute idealism, this ultimate is God, who is also equivalently termed 'The Absolute.' In such monistic schemes, the ultimate is illegitimately allowed a final, 'eminent' reality, beyond that ascribed to any of its accidents.

'Creativity' is another rendering of the Aristotelian 'matter,' and of the modern 'neutral stuff.' But it is divested of the notion of passive receptivity, either of 'form,' or of external relations; it is the pure notion of the activity conditioned by the objective immortality of the actual world—a world which is never the same twice, though always with the stable element of divine ordering. Creativity is without a character of its own in exactly the same sense in which the Aristotelian 'matter' is without a character of its own. It is that ultimate notion of the highest generality at the base of actuality. It cannot be characterized, because all characters are more special than itself. But creativity is always found under conditions, and described as conditioned. The non-temporal act of all-inclusive unfettered valuation [i.e., God] is at once a creature of creativity and a condition for creativity. It shares this double character with all creatures.

An actual entity feels as it does feel in order to be the actual entity which it is. In this way an actual entity satisfies Spinoza's notion of substance: it is *causa sui*. All actual entities share with God this characteristic of self-causation. For this reason every actual entity also shares with God the characteristic of transcending all other actual entities, including God. The universe is thus a creative advance into novelty. The alternative to this doctrine is a static morphological universe.

Two considerations of prime importance emerge from this initial statement. In the first place, creativity is not to be conceived as an "external agency with its own ulterior purposes"—i.e., it must not violate the ontological principle but must be explicable by an appeal to actual entities. Second, creativity is the concept that must account for the perpetual "creative advance into novelty" that is the cornerstone of Whitehead's process philosophy. The account of creativity in PR is terse to the point of obscurity. Whitehead's basic statements on the topic will now be presented, followed by an interpretive commentary.

'Creativity,' 'many,' 'one' are the ultimate notions involved in the meaning of the synonymous terms 'thing,' 'being,' 'entity.' The term 'many' presupposes the term 'one,' and the term 'one' presupposes the term 'many.' The term 'one' stands for the singularity of an entity. The term 'many' conveys the notion of 'disjunctive diversity'; this notion is an essential element in the concept of 'being.' There are many 'beings' in disjunctive diversity.

'Creativity' is [i] the principle of *novelty*. An actual occasion is a novel entity diverse from any entity in the 'many' which it unifies. Thus 'creativity' introduces novelty into the content of the many, which are the universe disjunctively. The creative action is the universe always becoming one in a particular unity of self-experience, and thereby adding to the multiplicity which is the universe as many.

'Creativity' is [ii] that ultimate principle by which the many, which are the universe disjunctively, become the one actual occasion, which is the universe conjunctively. It lies in the nature of things that the many enter into complex unity. In their natures, entities are disjunctively 'many' in process of passage into conjunctive unity. The fundamental inescapable fact is the creativity in virtue of which there can be no 'many things' which are not subordinated in a concrete unity. Thus a set of all actual occasions is by the nature of things a standpoint for another concrescence which elicits a concrete unity from those many actual occasions. It is inherent in the constitution of the immediate, present actuality that a future will supersede it. The creativity in virtue of which any relative complete actual world is, by the nature of things, the datum for a new concrescence, is termed 'transition.'

Thus the 'production of novel togetherness' is the ultimate notion embodied in the term 'concrescence.' The ultimate metaphysical principle is the advance from disjunction to conjunction, creating a novel entity other than the entities given in disjunction. The world expands through recurrent unifications of itself, each, by the addition of itself, automatically recreating the multiplicity anew. The novel entity is at once the togetherness of the 'many' which it finds, and also it is one among the disjunctive 'many' which it leaves; it is a novel entity, disjunctively among the many entities which it synthesizes. The many become one, and are increased by one.

Nature is never complete. It is always passing beyond itself. This is the creative advance of nature. The 'creative advance' is the application of this ultimate principle of creativity to each novel situation which it originates. The creative process is rhythmic: it swings from the publicity of many things to the individual privacy; and it swings back from the private individual to the publicity of the objectified individual. The former swing is dominated by the final cause which is the ideal; and the latter swing is dominated by the efficient cause which is actual. The oneness of the universe, and the oneness of each element in the universe, repeat themselves to the crack of doom in the creative advance from creature to creature.

The basic principle underlying these tightly packed sentences is enunciated twice by Whitehead: "It lies in the nature of things that the many enter into complex unity," and again, "The fundamental inescapable fact is the creativity in virtue of which there can be no 'many things' which are not subordinated in a concrete unity." In short, the universe abhors a "many" and moves, via the unity of a fresh concrescence, to overcome a "many." The thrust of the system is immediately evident when it is seen, however, that creativity is also "the principle of novelty." This means that the new unity that "subordinates" a "many" is itself a novel entity disjunctively diverse from everything else in the universe, so that in "subordinating" a "many" it itself in effect creates another "many" requiring "subordination." In short, to assuage the abhorrent situation is to re-create that very same abhorrent condition.

In this perpetual sequence is contained the basic rhythm of process. In Modes of Thought (p. 120) Whitehead writes: "There is a rhythm of process whereby creation produces natural pulsation, each pulsation forming a natural unit of historic fact." The natural units of historic fact are the actual entities and the rhythm is the alternation between "one" and "many" that repeats itself "to the crack of doom in the creative advance from creature to creature."

In this account of creativity there has been no repudiation of the ontological principle. Each individual among the "many" and each "one" that emerges are all alike actual occasions, whereas "creativity" is only the "universal of universals characterizing ultimate matter of fact" (PR, p. 31), the ultimate principle descriptive of the nature of actual entities. As the ultimate principle descriptive of the one-many relationship inhering in the coming-to-be of actual entities, creativity points up the fact that actual entities are not independent of and separate from one another. There is a perpetual advance to fresh actual occasions precisely because the actual entities are not wholly independent, but rather are linked in the creative process resulting from the one-many relationship that binds them together. This doctrine of creativity is therefore the natural outcome of two principles introduced in Chapter One that also emphasize the interdependence of actual entities, namely, the Principle of Relativity and the principle that every actual entity is superject as well as subject.

Chapter Three

THE PHASES
OF CONCRESCENCE

I
Introductory Statement

The process of concrescence is divisible into an initial stage of many feelings, and a succession of subsequent phases of more complex feelings integrating the earlier simpler feelings, up to the satisfaction which is one complex unity of feeling. This is the 'genetic' analysis of the satisfaction. The actual entity is seen as a process; there is a growth from phase to phase; there are processes of integration and of reintegration.

This genetic passage from phase to phase is not in physical time: the exactly converse point of view expresses the relationship of concrescence to physical time. The actual entity is the enjoyment of a certain quantum of physical time. But the genetic process is not the temporal succession: such a view is exactly what is denied by the epochal theory of time. Each phase in the genetic process presupposes the entire quantum, and so does each feeling in each phase. The subjective unity dominating the process forbids the division of that extensive quantum which originates with the primary phase of the subjective aim. It can be put shortly by saying, that physical time expresses some features of the growth, but *not* the growth of the features.

The analysis of an actual entity is only intellectual. Each actual entity is a cell with atomic unity. But in analysis it can only be understood as a process; it can only be felt as a process, that is to say, as in passage. The actual entity is divisible; but is in fact undivided. The authority of William James can be quoted in support of this conclusion. He writes: "Either your experience is of no content, of no change, or it is of a perceptible amount of content or change. Your acquaintance with reality grows literally by buds or drops of perception. Intellectually and on reflection you can divide these into components, but as immediately given, they come totally or not at all." (Cf. *Some Problems of Philosophy*, Ch. X.)

The prehensions in disjuction are abstractions; each of them is its subject viewed in that abstract objectification. The actuality is the totality of prehensions with subjective unity in process of concrescence into concrete unity.

There are an indefinite number of prehensions, overlapping, subdividing, and supplementary to each other. The principle, according to which a prehension can be discovered, is to take any component in the objective datum of the satisfaction; in the complex pattern of the subjective form of the satisfaction there will be a component with direct relevance to this element in the datum. Then in the satisfaction, there is a prehension of this component of the objective datum with that component of the total subjective form as its subjective form.

The genetic growth of this prehension can then be traced by considering the transmission of the various elements of the datum from the actual world, and—in the case of eternal objects—their origination in the conceptual prehensions. There is then a growth of prehensions, with integrations, eliminations, and determination of subjective forms. But the determination of successive phases of subjective forms, whereby the integrations have the characters that they do have, depends on the unity of the subject imposing a mutual sensitivity upon the prehensions. Thus a prehension, considered genetically, can never free itself from the incurable atomicity of the actual entity to which it belongs. The selection of a subordinate prehension from the satisfaction—as described above—

involves a hypothetical, propositional point of view. The fact is the satisfaction as one. There is some arbitrariness in taking a component from the datum with a component from the subjective form, and in considering them, on the ground of congruity, as forming a subordinate prehension. The justification is that the genetic process can be thereby analysed. This genetic process has now to be traced in its main outlines.

This preliminary statement (1) indicates the scope of the present chapter, (2) acknowledges a difficulty inherent in the notion of phases of concrescence, and (3) introduces considerations designed to mitigate the difficulty.

The chapter provides a genetic analysis of the satisfaction of actual entities—that is, it analyzes the phases of concrescence. The difficulty involves the relation of the genetic process to physical time. Whitehead incorporates the relativity theory of modern physics (discussed in Chapter Five, Section II) into the basic principles of his system. This theory entails the idea that there is no absolute time as a sort of container within which actual entities become; rather, time is an abstraction from the succession of actual entities. Concrescence is not in time; rather, time is in concrescence in the sense of being an abstraction from actual entities. But if this is the case, then it seems strange to speak of one phase of concrescence as prior to another when the passage from phase to phase is not in physical time. This is the difficulty.

William Christian (An Interpretation of Whitehead's Metaphysics, Yale, 1959, pp. 80–81) asks what sort of priority this priority of one phase to another could be, concluding that it is not temporal priority, logical priority, part-whole priority nor dialectical priority, but must be a priority sui generis.

It may seem to many that this suggestion is ad hoc and unsatisfactory. But a better suggestion is hard to find. This difficulty does not admit of easy solution and Whitehead scholars increasingly are going to have to puzzle over it, for the sequence of phases of concrescence is too central to the system to be abandoned, and yet difficult to incorporate into the system.

The final four paragraphs of the preceding introductory state-
ment introduce considerations that, it would seem, were thought by
Whitehead to mitigate the difficulty. Whether in fact they do is
open to question. Whitehead's point is that the analysis into phases
of concrescence is "only intellectual" and that there is "some arbi-
trariness" in distinguishing individual prehensions. The reader is
encouraged to ask himself, in the light of his reading of this chapter,
whether Whitehead is entitled to these last points, or whether he
puts such weight on the separation and sequence of prehensions that
it will not do to say that the genetic analysis is "only intellectual."

In a process of concrescence, there is a succession of phases in
which new prehensions arise by integration of prehensions in antecedent phases. The process continues till all prehensions are components in the one determinate integral satisfaction. There are
three successive phases of feelings, namely, a phase of 'conformal'
feelings, one of 'conceptual' feelings, and one of 'comparative' feelings. The two latter phases can be put together as the 'supplemental'
phase. The stages of comparison involve comparisons, and comparisons of comparisons.

This description of the phases of concrescence can be diagramed
in a way that simplifies much of the following discussion. The three
phases mentioned by Whitehead are more conveniently represented
as four. The fourth stage is obtained by breaking the third phase,
that of comparative feelings, into two stages; this split is justified
by the last sentence, which distinguishes between (1) comparisons
and (2) comparisons of comparisons. The third phase then becomes
the phase of simple *comparative feelings and the fourth phase be-*
comes the phase of complex *comparative feelings, frequently re-*
ferred to as intellectual feelings. Broken down this way, phases two
through four constitute the supplemental *phase, which stands, as*
the source of novelty, over against the primary, conformal, receptive,
or responsive phase, the phase of the given, or of the datum. These
relationships are schematized in Figure 2. There is more in Figure
2 than these relationships, however, and that "more" will be ex-
plicated in succeeding discussions.

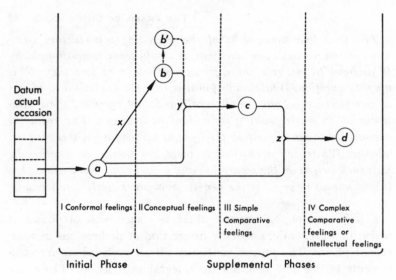

Figure 2. The Phases of Concrescence.

II

Phase I—Conformal Feelings

The primary stage in the concrescence of an actual entity is the way in which the antecedent universe enters into the constitution of the entity in question, so as to constitute the basis of its nascent individuality. A simple physical feeling is one feeling which feels another feeling. But the feeling felt has a subject diverse from the subject of the feeling which feels it. A multiplicity of simple physical feelings constitutes the first phase in the concrescence of the actual entity which is the common subject of all these feelings. All the more complex kinds of physical feelings arise in subsequent phases of concrescence, in virtue of integrations of simple physical feelings with each other and with conceptual feelings.

The first phase is the phase of pure reception of the actual world in its guise of objective datum for aesthetic synthesis. This datum, which is the primary phase in the process constituting an actual entity, is nothing else than the actual world itself in its character of a possibility for the process of being felt. This exemplifies the metaphysical principle that every 'being' is a potential for a 'becoming.' The actual world is the 'objective content' of each new creation.

A simple physical feeling has the dual character of being the cause's feeling re-enacted for the effect as subject. But this transference of feeling effects a partial identification of cause with effect, and not a mere representation of the cause. It is the cumulation of the universe and not a stage-play about it. By reason of this duplicity in a simple feeling there is a vector character which transfers the cause into the effect. It is a feeling *from* the cause which acquires the subjectivity of the new effect without loss of its original subjectivity in the cause.[1] This primary phase of simple physical feelings constitutes the machinery by reason of which the creativity transcends the world already actual, and yet remains conditioned by that actual world in its new impersonation.

Simple physical feelings embody the reproductive character of nature, and also the objective immortality of the past. In virtue of these feelings time is the conformation of the immediate present to the past. Such feelings are 'conformal' feelings. The conformal stage merely transforms the objective content into subjective feelings.

1.   Three Categoreal Obligations

Concrescence, the process of integrating the initial welter of many simple physical feelings into the one complex unity of feeling that is the satisfaction, proceeds in accordance with several "Categoreal Obligations," or "Categories of Obligation." Various categories are operative at various phases of concrescence. Already in the first phase three categoreal conditions are operative, and these must now be introduced. Others will be presented as they become relevant.

The status of the Categories of Obligation will become clearer as the discussion proceeds. It has been suggested that they are similar to Kant's Categories, except that they operate throughout the world

---

[1] The reader may wish to refer to Section II of Chapter One, where a simple physical feeling has been described in detail.

instead of merely in the experience of conscious persons. They are conditions of all possible actual worlds, rather than merely conditions of all possible experience. That is, because in Whitehead's system experience is not restricted to conscious experience, but in a primitive form permeates all reality as prehensive activity, Whitehead's Categories of Obligation structure all reality and not merely conscious human experience.

There are three main categoreal conditions which flow from the final nature of things. These three conditions are: (i) the category of subjective unity, (ii) the category of objective identity, and (iii) the category of objective diversity. Later we shall isolate other categoreal conditions. But the three conditions mentioned above have an air of ultimate metaphysical generality.

CATEGORY I. The many feelings which belong to an incomplete phase in the process of an actual entity, though unintegrated by reason of the incompleteness of the phase, are compatible for synthesis by reason of the unity of their subject.

This is the category of 'subjective unity.' This category is one expression of the general principle that the one subject is the final end which conditions each component feeling. Although in any incomplete phase there are many unsynthesized feelings, yet each of these feelings is conditioned by the other feelings. The process of each feeling is such as to render that feeling integrable with the other feelings.

The novel actual entity, which is the effect, is the reproduction of the many actual entities of the past. But in this reproduction there is abstraction from their various totalities of feeling. This abstraction is required by the categoreal conditions for compatible synthesis in the novel unity. The limitation, whereby the actual entities felt are severally reduced to the perspective[2] of one of their own feelings, is imposed by the categoreal condition of subjective unity, requiring a harmonious compatibility in the feelings of each incomplete phase. This subjective insistence on consistency may, from the beginning, replace positive feelings by negative prehensions. Feelings are dis-

[2] For the discussion of "perspective," see Chapter One, Section II, Sub-section 1.

missed by negative prehensions, owing to their lack of compliance with categoreal demands.

This mutual determination of the elements involved in a feeling is one expression of the truth that the subject of the feeling is *causa sui*. The partial nature of a feeling, other than the complete satisfaction, is manifest by the impossibility of understanding its generation without recourse to the whole subject. There is a mutual sensitivity of feelings in one subject, governed by categoreal conditions. This mutual sensitivity expresses the notion of final causation in the guise of a pre-established harmony.

CATEGORY II. There can be no duplication of any element in the objective datum of the satisfaction of an actual entity, so far as concerns the function of that element in that satisfaction.

This is the 'category of objective identity.' This category asserts the essential self-identity of any entity as regards its status in each individualization of the universe. In such a concrescence one thing has one rôle, and cannot assume any duplicity. This is the very meaning of self-identity, that, in any actual confrontation of thing with thing, one thing cannot confront itself in alien rôles. Any one thing remains obstinately itself playing a part with self-consistent unity. This category is one ground of incompatibility.

CATEGORY III. There can be no 'coalescence' of diverse elements in the objective datum of an actual entity, so far as concerns the functions of those elements in that satisfaction.

This is the 'category of objective diversity.' Here the term 'coalescence' means the self-contradictory notion of diverse elements exercising an absolute identity of function, devoid of the contrasts inherent in their diversities. In other words, in a real complex unity each particular component imposes its own particularity on its status. No entity can have an abstract status in a real unity. Its status must be such that only it can fill and only that actuality can supply.

2. *Three Categoreal Obligations Illustrated*

The importance of these categories can only be understood by considering each actual world in the light of a 'medium' leading up to the concrescence of the actual entity in question. Any actual entity, which we will name A, feels other actual entities which we will name B, C, and D. Thus B, C, and D all lie in the actual world of A. But C and D may lie in the actual world of B, and are then felt by it; also D may lie in the actual world of C and be felt by it. This example might be simplified, or might be changed to one of any degree of complication. Now B, as an initial datum for A's feeling, also presents C and D for A to feel through its mediation. Also C, as an initial datum for A's feeling, also presents D for A to feel through its mediation. Thus, in this artificially simplified example, A has D presented for feeling through three distinct sources: (i) directly as a crude datum, (ii) by the mediation of B, and (iii) by the mediation of C. This threefold presentation is D, in its function of an initial datum for A's feeling of it, so far as concerns the mediation of B and C. But, of course, the artificial simplification of the medium to two intermediaries is very far from any real case. The medium between D and A consists of all those actual entities which lie in the actual world of A and not in the actual world of D. For the sake of simplicity the explanation will continue in terms of this threefold presentation.

There are thus three sources of feeling, D direct, D in its nexus with C, and D in its nexus with B. Thus in the basic phase of A's concrescence there arise three prehensions of the datum D. According to the first category these prehensions are not independent. This subjective unity of the concrescence introduces negative prehensions, so that D in the direct feeling is not felt in its formal completeness, but objectified with the elimination of such of its prehensions as are inconsistent with D felt through the mediation of B, and through the mediation of C. Thus the three component feelings of the first phase are consistent, so as to pass into the integration of the second phase in which there is A's one feeling of a coherent objectification of D. Since D is necessarily self-consistent, the inconsistencies must arise from the subjective forms of the prehensions of D by B directly, by C directly, and by A directly. These inconsistencies lead to the eliminations in A's total prehension of D.

In this process, the negative prehensions which effect the elimination are not merely negligible. The process through which a feeling passes in constituting itself, also records itself in the subjective form of the integral feeling. The negative prehensions have their own subjective forms which they contribute to the process. A feeling bears on itself the scars of its birth: it recollects as a subjective emotion its struggle for existence; it retains the impress of what it might have been, but is not. It is for this reason that what an actual entity has avoided as a datum for feeling may yet be an important part of its equipment. The actual cannot be reduced to mere matter of fact in divorce from the potential.

Only the first category has been explicitly alluded to. It must now be pointed out how the other categories have been tacitly presupposed.

The fact that there is integration at all arises from the condition expressed by the category of objective identity. The same entity, be it actual entity or be it eternal object, cannot be felt twice in the formal constitution of one concrescence. The incomplete phases with their many feelings of one object are only to be interpreted in terms of the final satisfaction with its one feeling of that one object. Thus objective identity requires integration of the many feelings of one object into the one feeling of that object.

The third category is concerned with the antithesis to oneness, namely, diversity. An actual entity is not merely one; it is also definitely complex. But, to be definitely complex is to include definite diverse elements in definite ways. The category of objective diversity expresses the inexorable condition—that a complex unity must provide for each of its components a real diversity of status, with a reality which bears the same sense as its own reality and is peculiar to itself. In other words, a real unity cannot provide sham diversities of status for its diverse components.

This category is in truth only a particular application of the second category. For a 'status' is after all *something;* and, according to the category of objective identity, it cannot duplicate its rôle. Thus if the 'status' be the status of *this,* it cannot in the same sense be the status of *that.*

Thus the process of integration, which lies at the very heart of the concrescence, is the urge imposed on the concrescent unity of that universe by the three categories of subjective unity, of objective identity, and of objective diversity.

III

Phase II—Conceptual Feelings

In each concrescence there is a twofold aspect of the creative urge. In one aspect there is the origination of simple causal feelings; and in the other aspect there is the origination of conceptual feelings.[3] These contrasted aspects will be called the physical and the mental poles of an actual entity. No actual entity is devoid of either pole; though their relative importance differs in different actual entities. The integration of the physical and mental side into a unity of experience is a self-formation which is a process of concrescence. Thus an actual entity is essentially dipolar, with its physical and mental poles; and even the physical world cannot be properly understood without reference to its other side, which is the complex of mental operations.

The physical inheritance is essentially accompanied by a conceptual reaction. We have to discuss the categoreal conditions for such derivation of conceptual feelings from the physical feelings relating to the temporal world. By the categoreal condition of subjective unity—Category I—the initial phase of physical feelings has the unity of feelings compatible for integration into one feeling of the actual world. But the completed determination of the subjective form of this final 'satisfaction' awaits the origination of conceptual feelings whose subjective forms introduce the factor of 'valuation,' that is, 'valuation up' *or* 'valuation down.'

Thus a supplementary phase succeeds to the initial purely physical phase. This supplementary phase starts with two subordinate phases of conceptual origination, and then passes into phases of integration, and of reintegration.

This last sentence can be superimposed upon Figure 2. The initial "two subordinate phases" of supplementation both occur in Phase II of the diagram; they are respectively indicated by the circles b *and* b′. *The phase of "integration" refers to Phase III of*

---

3 See Section I of Chapter Two, where conceptual feelings are discussed. It will be recalled that a conceptual feeling "is a feeling whose 'datum' is an eternal object," whereas causal, or physical feelings are "prehensions of actual entities."

the diagram and the phase of reintegration to Phase IV. The re-
mainder of this section elucidates the two subordinate stages of
Phase II.

[Given that a conceptual feeling is a feeling whose datum is an
eternal object,] the initial problem is to discover the principles according to which some eternal objects are prehended positively and
others are prehended negatively. Some are felt and others are eliminated. In the solution of this problem additional categoreal conditions must be added to the three such conditions which have already
been explained. The conditions have regard to the origination, and
coordination, of conceptual feelings.

CATEGORY IV. THE CATEGORY OF CONCEPTUAL VALUATION. From
each physical feeling there is the derivation of a purely conceptual
feeling whose datum is the eternal object exemplified in the definiteness of the actual entity, or the nexus, physically felt.[4]

This category maintains the old principle that mentality originates from sensitive experience. It lays down the [new] principle
that all sensitive experience originates mental operations. It does
not, however, mean that there is no origination of other mental
operations derivative from these primary mental operations. Nor does
it mean that these mental operations involve consciousness, which is
the product of intricate integration [in later phases].

The mental pole originates as the conceptual counterpart of
operations in the physical pole. The two poles are inseparable in
their origination. The mental pole starts with the conceptual regis-
tration of the physical pole.

1. Conceptual Reversion

CATEGORY V. THE CATEGORY OF CONCEPTUAL REVERSION. There
is secondary origination of conceptual feelings with data which are

---

[4] In terms of Figure 2 this means that the conceptual feeling represented by
circle *b* in Phase II is derived as indicated by arrow *x*, from the simple physical
feeling represented by circle *a* in Phase I. In the simple physical feeling at *a*
certain eternal objects are felt as being immanent, as being *realized* determinants
of the datum; at *b* these same eternal objects are felt conceptually—i.e., they
are "pried out" of their immanence and felt as transcendent, as *a capacity for*
determination. See Section I of Chapter Two for a discussion of eternal objects
as immanent and transcendent.

partially identical with, and partially diverse from, the eternal objects forming the data in the primary phase of the mental pole; the determination of identity and diversity depending on the subjective aim at attaining depth of intensity by reason of contrast.

Note that Category IV concerns conceptual reproduction of physical feeling, and Category V concerns conceptual diversity from physical feeling. Thus the first [sub-] phase of the mental pole [*b* of Figure 2] is conceptual reproduction, and the second [sub-] phase [*b'*] is a phase of conceptual reversion. In this second [sub-] phase the proximate novelties are conceptually felt. This is the process by which the subsequent enrichment of subjective forms, both in qualitative pattern, and in intensity through contrast, is made possible by the positive conceptual prehension of relevant alternatives. There is a conceptual contrast of physical incompatibles. This is the category which, as thus stated, seems to limit the rigid application of Hume's principle [of the derivation of conceptual experience from physical experience]. Indeed Hume himself admitted exceptions.[5] It is the category by which novelty enters the world; so that even amid stability there is never undifferentiated endurance. But, as the category states, reversion is always limited by the necessary inclusion of elements identical with elements in feelings of the antecedent phase.

The question, how, and in what sense, one unrealized eternal object can be more, or less, proximate to an eternal object in realized ingression—that is to say, in comparison with any other unfelt eternal object—is left unanswered by this category of reversion. In conformity with the ontological principle, this question can be answered only by reference to some actual entity. Every eternal object has entered into the conceptual feelings of God. Thus, a more fundamental account must ascribe the reverted conceptual feeling in a temporal subject to its conceptual feeling derived, according to Category IV, from the hybrid physical feeling of the relevancies conceptually ordered in God's experience.

[5] For example, the famous instance of the missing shade of blue mentioned in the first section of the *Treatise*. To imagine a particular shade of blue never actually experienced, which Hume admits to be possible, would be an instance of a reverted conceptual feeling.

2. Reversion, God, and Subjective Aim

The following, "more fundamental" account of conceptual reversion is also a more thorough discussion of the relation between God and subjective aim, a topic that was sketched in Chapter Two. To begin, a hybrid physical feeling is mentioned in the last sentence of the preceding section, and this notion requires an explanation.

A feeling will be called 'physical' when its datum involves objectifications of other actual entities. The special case of 'simple physical feelings' was discussed. A feeling belonging to this special case has as its datum only one actual entity, and this actual entity is objectified by one of its feelings. Such feelings are subdivided into 'pure physical feelings' and 'hybrid physical feelings.' In a 'pure physical feeling' the actual entity which is the datum is objectified by one of its own physical feelings. In a 'hybrid physical feeling' the actual entity forming the datum is objectified by one of its own conceptual feelings. A hybrid physical feeling originates for its subject a conceptual feeling with the same datum as that of the conceptual feeling of the antecedent subject. But the two conceptual feelings in the two subjects respectively may have different subjective forms. There are two sub-species of hybrid feelings: (i) those which feel the conceptual feelings of temporal actual entities, and (ii) those which feel the conceptual feelings of God.

The objectification of God in a temporal subject is effected by the hybrid feelings with God's conceptual feelings as data. Those of God's feelings which are positively prehended are those with some compatibility of contrast, or of identity, with physical feelings transmitted from the temporal world. Thus the primary phase [of an actual entity contains] a hybrid physical feeling of God, in respect to God's conceptual feeling which is immediately relevant to the universe 'given' for that concrescence. There is then, according to the category of conceptual valuation, i.e. Categoreal Obligation IV, a derived conceptual feeling which reproduces for the subject the data and valuation of God's conceptual feeling. This conceptual feeling is the initial conceptual aim [i.e., the subjective aim] referred to in the preceding [chapter].

It will help to superimpose this conclusion on Figure 2. What Whitehead has said in this final paragraph is that whenever there is a reversion—i.e., a b′ *circle in Phase II—there is presupposed in Phase I a hybrid physical feeling of God from which that reversion is derivable by Category IV, the Category of Conceptual Valuation. This entails that reversion as a category is superfluous, since any feeling resulting from reversion can be explained equally well in terms of God and Category IV. But though superfluous, reversion is convenient and is retained because there is a close relationship between any* b *and its related* b′. *The fundamental explanation for this is given in the following sentence: "Those of God's feelings which are positively prehended are those with some compatibility of contrast, or of identity, with physical feelings transmitted from the temporal world." To use the Category of Reversion is to emphasize this "compatibility of contrast, or of identity" between conceptual novelties and inherited fact. These points are summarized in the following passage.*

With this amplification the doctrine, that the primary phase of a temporal actual entity is physical, is recovered. When we take God into account, then we can assert without any qualification Hume's principle, that all conceptual feelings are derived from physical feelings. In this way, by the recognition of God's characterization of the creative act, a more complete rational explanation is attained. The category of reversion is then abolished; and Hume's principle of the derivation of conceptual experience from physical experience remains without any exception. The limitation of Hume's principle introduced by the consideration of the Category of Conceptual Reversion is to be construed as referring merely to the transmission from the temporal world, leaving God out of account.

3. *Valuations and Further Categoreal Obligations*

The subjective form of a feeling is how that feeling feels its datum; see Chapter One, Section II. In the case of conceptual feelings, the subjective form is therefore how the datum eternal object is felt. "Valuation" is the name for this "how" of feeling in the case of conceptual feelings. Valuation is either valuation up or valuation down, adversion or aversion. Valuation looks ahead to the third phase of concrescence since it establishes the importance of its datum eternal objects for subsequent phases of feeling.

The subjective form of a conceptual feeling has the character of a 'valuation,' and this notion must now be explained. If in the conceptual feelings there is valuation upward, this is 'adversion.' But if in the conceptual feelings there is valuation downward, this is 'aversion.' Thus 'adversion' and 'aversion' are types of 'decision.' There is an autonomy in the formation of the subjective forms of conceptual feelings, conditioned only by the unity of the subject.

A valuation has three characteristics:

(i) The valuation is dependent on the other feelings in its phase of origination.

(ii) The valuation determines in what status the eternal object has ingression into [subsequent feeling].

(iii) The valuation values up, or down, so as to determine the intensive importance accorded to the [datum] eternal object by the subjective form of [subsequent] feeling.

Thus, according as the valuation of the conceptual feeling is a 'valuation up' or a 'valuation down,' the importance of the [datum] eternal object as felt in [subsequent phases of] feeling is enhanced, or attenuated.

This discussion of valuation leads naturally to two further Cate-goreal Obligations, Category VII, the Category of Subjective Harmony, and Category VIII, the Category of Subjective Intensity. (Note that Category VI has been left out for the moment; it will be introduced in Chapter Four.) In the paragraphs that follow, the notion of a "contrast" will be introduced. A contrast is explained by Whitehead as follows: "Whatever is a datum for a feeling has a unity as felt. Thus the many components of a complex datum have a unity: this unity is a 'contrast' of entities." The third and fourth phases of concrescence are phases of "integration and reintegration." Feelings in these phases have complex groupings of earlier feelings in the concrescence as their data; this is to say that feelings in these phases have "contrasts" as data.

The seventh categoreal condition governs the efficacy of conceptual feelings both in the completion of their own subjects and also in the objectifications of their subjects in subsequent concrescence. It is the category of 'subjective harmony.'

CATEGORY VII. THE CATEGORY OF SUBJECTIVE HARMONY. The valuations of conceptual feelings are mutually determined by their adaptation to be joint elements in a satisfaction aimed at by the subject, [i.e.,] by the adaptation of those feelings to be contrasted elements congruent with the subjective aim. This categoreal condition should be compared with the category of 'subjective unity,' and also with the category of 'conceptual reversion.'

Category I and Category VII jointly express a pre-established harmony in the process of concrescence of any one subject. Category I has to do with data felt, and Category VII with the subjective forms of the conceptual feelings. This pre-established harmony is an outcome of the fact that no prehension can be considered in abstraction from its subject, although it originates in the process creative of its subject.

By the Category of Subjective Unity, and by the seventh category of Subjective Harmony, all origination of feelings is governed by the subjective imposition of aptitude for final synthesis. In the former category the intrinsic inconsistencies, termed 'logical,' are the formative conditions in the pre-established harmony. In this seventh category, and in the Category of Reversion, aesthetic adaptation for an end is the formative condition in the pre-established harmony. These three categories express the ultimate particularity of feelings. For the superject which is their outcome is also the subject which is operative in their production. They are the creation of their own creature. It is only by reason of the Categories of Subjective Unity, and of Subjective Harmony, that the process constitutes the character of the product, and that conversely the analysis of the product discloses the process. The point to be noticed is that the actual entity, in a state of process during which it is not fully definite, determines its own ultimate definiteness. This is the whole point of moral responsibility. Such responsibility is conditioned by the limits of the data, and by the categoreal conditions of concrescence.

But autonomy is negligible unless the complexity is such that there is great energy in the production of conceptual feelings according to the Category of Reversion. Reversions are the conceptions which arise by reason of the lure of contrast, as a condition for intensity of experience. This lure is expressible as a categoreal condition.

CATEGOREAL CONDITION VIII. THE CATEGORY OF SUBJECTIVE INTENSITY. The subjective aim, whereby there is origination of conceptual feeling, is at intensity of feeling (a) in the immediate subject, and (β) in the relevant future.

This double aim—at the *immediate* present and the *relevant* future—is less divided than appears on the surface. For the determination of the *relevant* future, and the *anticipatory* feeling respecting provision for its grade of intensity, are elements affecting the immediate complex of feeling. The greater part of morality hinges on the determination of relevance in the future. The relevant future consists of those elements in the anticipated future which are felt with effective intensity by the present subject by reason of the real potentiality for them to be derived from itself.

We first note (i) that intensity of feeling due to any realized ingression of an eternal object is heightened when that eternal object is one element in a realized contrast between eternal objects, and (ii) that two or more contrasts may be incompatible for joint ingression, *or* may jointly enter into a higher contrast.

It follows that balanced complexity is the outcome of this final category of subjective aim. Here 'complexity' means the realization of contrasts, of contrasts of contrasts, and so on; and 'balance' means the absence of attenuations due to the elimination of contrasts which some elements in the pattern would introduce and other elements inhibit. Unless there is complexity, ideal diversities lead to physical impossibilities, and thence to impoverishment. It requires a complex constitution to stage diversities as consistent contrasts.

The contrasts produced by reversion are contrasts required for the fulfilment of the aesthetic ideal. Thus there is the urge towards the realization of the maximum number of eternal objects subject to the restraint that they must be under conditions of contrast. But this limitation to 'conditions of contrast' is the demand for 'balance.' For 'balance' here means that no realized eternal object shall eliminate potential contrasts between other realized eternal objects. Such eliminations attenuate the intensities of feeling derivable from the ingressions of the various elements of the pattern. Thus so far as the immediate present subject is concerned, the origination of conceptual valuation according to Category IV is devoted to such a disposition of emphasis as to maximize the integral intensity derivable from the most favourable balance. The subjective aim is the selection of the balance amid the given materials.

But one element in the immediate feelings of the concrescent subject is comprised of the anticipatory feelings of the transcendent future in its relation to immediate fact. This is the feeling of the objective immortality inherent in the nature of actuality. Such anticipatory feelings involve realization of the relevance of eternal objects as decided in the primordial nature of God. In so far as these feelings in the higher organisms rise to important intensities there are effective feelings of the more remote alternative possibilities. Such feelings are the conceptual feelings which arise in accordance with the Category of Reversion.

But there must be 'balance,' and 'balance' is the adjustment of identities and diversities for the introduction of contrast with the avoidance of inhibitions by incompatibilities. Thus this secondary [sub-] phase, involving the future, introduces reversion and is subject to this final category. Each reverted conceptual feeling [*b'*] has its datum largely identical with that of its correlate primary feeling [its correlate *b*]. In this way, readiness for synthesis is promoted. But the introduction of contrast is obtained by the differences, or reversions, in some elements of the complex data. The Category [of Subjective Intensity] expresses the rule that what is identical, and what is reverted, are determined by the aim at a favourable balance. The reversion is due to the aim at complexity as one condition for intensity.

IV

Phase III—Simple Comparative Feelings

A simple comparative feeling is a feeling that compares, or holds in the unity of a contrast, a simple physical feeling from Phase I and a conceptual feeling from Phase II (normally the conceptual counterpart of the physical feeling, derived from it by conceptual valuation or conceptual reversion). In Figure 2, circle c represents a simple comparative feeling and bracket y represents the datum for c, which is a contrast unifying a and b (or b'). Whitehead sometimes *refers to this contrast as the "integrated datum" for c, and this leads him occasionally to refer to c as an "integral comparative feeling."*

Simple comparative feelings are of two main types; they are either physical purposes or propositional feelings. Physical purposes tend to be terminal—i.e., they occur in primitive actual occasions and inhibit further integrations. Propositional feelings, on the other hand, are lures for further integration and appear at the third phase of sophisticated actual entities that will subsequently push on to the fourth phase of complex comparative feelings. The two species of physical purposes will be presented first, then propositions and propositional feelings will be discussed, and third a brief section will explicitly consider the difference between physical purposes and propositional feelings.

1. Physical Purposes

Conceptual feelings and simple causal feelings constitute the two main species of 'primary' feelings. All other feelings of whatever complexity arise out of a process of integration which starts with a phase of these primary feelings. The integration of each simple physical feeling [from Phase I] with its conceptual counterpart [from Phase II] produces in a subsequent phase [III] a physical feeling whose subjective form of re-enaction has gained or lost subjective intensity according to the valuation up, or the valuation down, in the conceptual feeling. This is the phase of physical purpose.

In the integral comparative feeling the datum is the contrast of the conceptual datum with the reality of the objectified nexus. The physical feeling is feeling a real fact; the conceptual feeling is valuing an abstract possibility. The new datum is the compatibility or incompatibility of the fact as felt with the eternal object as a datum in feeling. This synthesis of a pure abstraction with a real fact, as in feeling, is a generic contrast. In respect to physical purposes, the cosmological scheme which is here being developed, requires us to hold that all actual entities include physical purposes. The constancy of physical purposes explains the persistence of the order of nature.

The chain of stages in which a physical purpose originates is: (i) there is a physical feeling; (ii) the primary conceptual correlate of the physical feeling is generated, according to categoreal condition IV; (iii) this physical feeling is integrated with its conceptual correlate to form the physical purpose. Such physical purposes are called physical purposes of the first species.

In such a physical purpose, the datum is the generic contrast between the nexus,[6] felt in the physical feeling, and the eternal object valued in the conceptual feeling. This eternal object is also exemplified as the pattern of the nexus. Thus the conceptual valuation now closes in upon the feeling of the nexus as it stands in the generic contrast, exemplifying the valued eternal object. This valuation accorded to the physical feeling endows the transcendent creativity with the character of adversion, or of aversion. The character of adversion secures the reproduction of the physical feeling, as one element in the objectification of the subject beyond itself. Such reproduction may be thwarted by incompatible objectification derived from other feelings. But a physical feeling, whose valuation produces adversion, is thereby an element with some force of persistence into the future beyond its own subject. It is felt and re-enacted down a route of occasions forming an enduring object.

When there is aversion, instead of adversion, the transcendent creativity assumes the character that it inhibits, or attenuates, the objectification of that subject in the guise of that feeling. Thus aversion tends to eliminate one possibility by which the subject may itself be objectified in the future. Thus adversions promote stability; and aversions promote change without any indication of the sort of change. In itself an aversion promotes the elimination of content, and the lapse into triviality.

Thus the conceptual feeling with its valuation has primarily the character of purpose, since it is the agent whereby the decision is

---

[6] A nexus is a unified collection of actual entities. Whereas a *simple* physical feeling feels a single actual occasion as datum, other physical feelings have as datum a nexus of actual occasions. In this latter case, the datum for the derived conceptual feeling is a complex eternal object, referred to as a pattern. The concept "nexus" is analyzed in detail in Chapter Four. *Nexūs* is the plural of *nexus*.

made as to the causal efficacy of its subject in its objectifications beyond itself. But it only achieves this character of purpose by its integration with the physical feeling from which it originates.

The bare character of mere responsive re-enaction constituting the original physical feeling in its first phase, is enriched in the second phase by the valuation accruing from integration with the conceptual correlate. In this way, the dipolar character of concrescent experience provides in the physical pole for the objective side of experience, derivative from an external actual world, and provides in the mental pole for the subjective side of experience, derivative from the subjective conceptual valuations correlate to the physical feelings. The mental operations have a double office. They achieve, in the immediate subject, the subjective aim of that subject as to the satisfaction to be obtained from its own initial data. In this way the decision derived from the actual world, which is the efficient cause, is completed by the decision embodied in the subjective aim, which is the final cause. Secondly, the physical purposes of a subject by their valuations determine the relative efficiency of the various feelings to enter into the objectifications of that subject in the creative advance beyond itself. In this function, the mental operations determine their subject in its character of an efficient cause. Thus the mental pole is the link whereby the creativity is endowed with the double character of final causation, and efficient causation. The mental pole is constituted by the decisions in virtue of which matters of fact enter into the character of the creativity.

According to this explanation, self-determination is always imaginative in its origin. The deterministic efficient causation is the inflow of the actual world in its own proper character of its own feelings, with their own intensive strength, felt and re-enacted by the novel concrescent subject. But this re-enaction has a mere character of conformation to pattern. The subjective valuation is the work of novel conceptual feeling; and in proportion to its importance, acquired in complex processes of integration and reintegration, this autonomous conceptual element modifies the subjective forms throughout the whole range of feeling in that concrescence and thereby guides the integrations.

In so far as there is negligible autonomous energy [and this is the case with physical purposes], the subject merely receives the physical feelings, confirms their valuations according to the 'order' of that epoch, and transmits by reason of its own objective immortality. Its own flash of autonomous individual experience is negligible. We have—in the case of the simpler actual entities—an example of the transference of energy in the physical world. So far as we can see, inorganic entities are vehicles for receiving, for storing in a napkin, and for restoring without loss or gain. But as soon as individual experience is not negligible [i.e., when the fourth phase of concrescence is reached], the autonomy of the subject in the modification of its initial subjective aim must be taken into account.

The second species of physical purposes is due to the origination of reversions in the mental pole. It is due to this second species that vibration and rhythm have a dominating importance in the physical world.

When [a] reverted conceptual feeling acquires a relatively high intensity of upward valuation in its subjective form, the resulting integration of physical feeling, primary conceptual feeling, and secondary conceptual feeling, produces a more complex physical purpose than in the former case when the reverted conceptual feeling was negligible. There is now the physical feeling as valued by its integration with the primary conceptual feeling, the integration with the contrasted secondary conceptual feeling, the heightening of the scale of subjective intensity by the introduction of conceptual contrast, and the concentration of this heightened intensity upon the reverted feeling in virtue of its being the novel factor introducing the contrast. The physical purpose thus provides the creativity with a complex character, which is governed by the Category of Conceptual Reversion, in virtue of which the secondary conceptual feeling arises; by the Category of Subjective Harmony, in virtue of which the subjective forms of the two conceptual feelings are adjusted to procure the subjective aim; and by the Category of Subjective Intensity, in virtue of which the aim is determined to the attainment of balanced intensity from feelings integrated in virtue of near-identity, and contrasted in virtue of reversions.

In the successive occasions of an enduring object in which the inheritance is governed by this complex physical purpose, the reverted conceptual feeling is transmitted into the next occasion as physical feeling, and the pattern of the original physical feeling now reappears as the datum in the reverted conceptual feeling. Thus there is a chain of contrasts in the physical feelings of the successive occasions. This chain is inherited as a vivid contrast of physical feelings, and in each occasion there is the physical feeling with its primary valuation in contrast with the reverted conceptual feeling. The formal constitutions of successive occasions are characterized by contraries supervening upon the aboriginal data, but with a regularity of alternation which procures stability in the life-history. Contrast is thus gained. In physical science, this is 'vibration.' This is the main character of the life-histories of an inorganic physical object, stabilized in type.

Thus an enduring object gains the enhanced intensity of feeling arising from contrast between inheritance and novel effect, and also gains the enhanced intensity arising from the combined inheritance of its stable rhythmic character throughout its life-history. It has the weight of repetition, the intensity of contrast, and the balance between the two factors of the contrast. In this way the association of endurance with rhythm and physical vibration is to be explained. They arise out of the conditions for intensity and stability. The subjective aim is seeking width with its contrasts, within the unity of a general design.

2. *Propositions and Propositional Feelings*

A proposition enters into experience as the entity forming the datum of a complex feeling derived from the integration of a physical feeling with a conceptual feeling. A propositional feeling is a feeling whose objective datum is a proposition.[7]

Now a conceptual feeling does not refer to *the* actual world, in the sense that the history of *this* actual world has any peculiar relevance to its datum. This datum is an eternal object; and an eternal object refers only to the purely general *any* among undetermined actual entities. In itself an eternal object evades any selection among actualities or epochs. You cannot know what is red by merely thinking of redness. You can only find red things by adventuring amid physi-

[7] These relationships can be visualized in terms of Figure 2. The proposition itself is represented by bracket *y*—the proposition is the integration of a physical feeling with a conceptual feeling. Circle *c* represents a propositional feeling that has as its datum the proposition represented by bracket *y*.

cal experiences in *this* actual world. This doctrine is the ultimate ground of empiricism; namely, that eternal objects tell no tales as to their ingressions.

But now a new kind of entity presents itself. Such entities are the tales that perhaps might be told about particular actualities. Such entities are neither actual entities, nor eternal objects, nor feelings. They are propositions. A proposition must be true or false. Herein a proposition differs from an eternal object; for no eternal object is ever true or false. This difference between propositions and eternal objects arises from the fact that truth and falsehood are always grounded upon a reason. But according to the ontological principle, a reason is always a reference to determinate actual entities. Now an eternal object, in itself, abstracts from all determinate actual entities, including even God. It is merely referent to *any* such entities, in the absolutely general sense of *any*. Then there can be no reason upon which to found the truth or falsehood of an eternal object.

But a proposition, while preserving the indeterminateness of an eternal object, makes an incomplete abstraction from determinate actual entities. It is a complex entity, with determinate actual entities among its components. These determinate actual entities, considered *formaliter* and not as in the abstraction of the proposition, do afford a reason determining the truth or falsehood of the proposition. But the proposition in itself, apart from recourse to these reasons, tells no tale about itself; and in this respect it is indeterminate like the eternal objects.

A propositional feeling (as has been stated) arises from a special type of integration synthesizing a physical feeling with a conceptual feeling. The objective datum of the physical feeling is either one actual entity, if the feeling be simple, or is a determinate nexus of actual entities, if the physical feeling be more complex. The datum of the conceptual feeling is an eternal object which is referent to any actual entities, where the *any* is absolutely general and devoid of selection. In the integrated objective datum the physical feeling provides its determinate set of actual entities, indicated by their felt physical relationships to the subject of the feeling. These actual entities are the logical subjects of the proposition. The absolute generality of the notion of *any*, inherent in an eternal object, is thus eliminated in the fusion. In the proposition, the eternal object, in respect to its possibilities as a determinant of nexūs, is restricted to these logical subjects. The proposition may have the restricted generality of referring to *any* among these provided logical subjects; or it may have the singularity of referring to the complete set of provided logical subjects as potential relata, each with its assigned status, in the complex pattern which is the eternal object. The proposition is the potentiality of the eternal object, as a determinant of definiteness, in some determinate mode of restricted reference to the logical subjects. This eternal object is the 'predicative pattern' of the proposition. The set of logical subjects is *either* completely singled out as *these* logical subjects in *this* predicative pattern *or* is collectively singled out as *any* of these logical subjects in *this* pattern, *or* as *some* of these logical subjects in *this* pattern.

Thus the physical feeling indicates the logical subjects and provides them respectively with that individual definition necessary to assign the hypothetic status of each in the predicative pattern. The conceptual feeling provides the predicative pattern. In this integration the two data are synthesized by a double elimination involving both data. The actual entities involved in the datum are reduced to a bare multiplicity in which each is a bare 'it' with the elimination of the eternal object really constituting the definiteness of that nexus. Thus in a proposition the logical subjects are reduced to the status of food for a possibility. Their real rôle in actuality is abstracted from; they are no longer factors in fact, except for the purpose of their physical indication. Each logical subject becomes a bare '*it*' among actualities, with *its* assigned hypothetical relevance to the predicate. But the integration rescues them from this mere multiplicity by placing them in the unity of a proposition with the given predicative pattern. Thus the actualities, which were first felt as sheer matter of fact, have been transformed into a set of logical subjects with the potentiality for realizing an assigned predicative pattern. The predicative pattern has also been limited by elimination. For as a datum in the conceptual feeling, it held its possibility for realization in respect to *absolutely any* actual entities; but in the proposition its possibilities are limited to *just these* logical subjects.

It is evident that the datum of the conceptual feeling reappears as the predicate in the proposition which is the datum of the integral, propositional feeling. In this synthesis the eternal object has suffered the elimination of its absolute generality of reference. The datum of the physical feeling has also suffered elimination. For the peculiar objectification of the actual entities, really effected in the physical feeling, is eliminated, except in so far as it is required for the services of the indication. The objectification remains only to indicate that definiteness which the logical subjects must have in order to be hypothetical food for that predicate. This necessary indication of the logical subjects requires the actual world as a systematic environment. For there can be no definite position in pure abstraction. The proposition is the possibility of *that* predicate applying in that assigned way to *those* logical subjects. In every proposition, as such and without going beyond it, there is complete indeterminateness so far as concerns its own realization in a propositional feeling, and as regards its own truth. The logical subjects are, nevertheless, in fact actual entities which are definite in their realized mutual relatedness. Thus the proposition is in fact true, or false. But its own truth, or its own falsity, is no business of a proposition. That question concerns only a subject entertaining a propositional feeling with that proposition for its datum. Such an actual entity is termed a 'prehending subject' of the proposition. An eternal object realized in respect to its pure potentiality as related to *determinate* logical subjects is termed a 'propositional feeling' in the mentality of the actual occasion in question.

A proposition, in abstraction from any particular actual entity which may be realizing it in feeling, is a manner of germaneness of a certain set of eternal objects to a certain set of actual entities. Every proposition presupposes those actual entities which are its logical subjects. It is a datum for feeling, awaiting a subject feeling it. Its relevance to the actual world by means of its logical subjects makes it a lure for feeling. In fact many subjects may feel it with diverse feelings, and with diverse sorts of feelings. The presupposed logical subjects may not be in the actual world of some actual entity. In this case, the proposition does not exist for that actual entity. The pure concept of *such* a proposition refers in the hypothetical future beyond that actual entity.

But according to the ontological principle, every proposition must be somewhere. The 'locus' of a proposition consists of those actual occasions whose actual worlds include the logical subjects of the proposition. When an actual entity belongs to the locus of a proposition, then conversely the proposition is an element in the lure for feeling of that actual entity. If by the decision of the concrescence, the proposition has been admitted into feeling, then the proposition constitutes *what* the feeling has felt. The proposition constitutes a lure for a member of its locus by reason of the germaneness of the complex predicate to the logical subjects, having regard to forms of definiteness in the actual world of that member, and to its antecedent phases of feeling.

There are two types of relationship between a proposition and the actual world of a member of its locus. The proposition may be conformal or non-conformal to the actual world, true or false.

When a conformal proposition is admitted into feeling, the reaction to the datum has simply resulted in the conformation of feeling to fact, with some emotional accession or diminution, by which the feelings inherent in alien fact are synthesized in a new individual valuation. The prehension of the proposition has abruptly emphasized one form of definiteness illustrated in fact.

When a non-conformal proposition is admitted into feeling, the reaction to the datum has resulted in the synthesis of fact with the alternative potentiality of the complex predicate. A novelty has emerged into creation. The novelty may promote or destroy order; it may be good or bad. But it is new, a new type of individual, and not merely a new intensity of individual feeling. That member of the locus has introduced a new form into the actual world; or at least, an old form in a new function.

The conception of propositions as merely material for judgments is fatal to any understanding of their rôle in the universe. In that purely logical aspect, non-conformal propositions are merely wrong, and therefore worse than useless. But in their primary rôle, they pave the way along which the world advances into novelty. Error is the price which we pay for progress.

The interest in logic, dominating overintellectualized philosophers, has obscured the main function of propositions in the nature of things. They are not primarily for belief, but for feeling at the physical level of unconsciousness. They constitute a source for the origination of feeling which is not tied down to mere datum. Unfortunately propositions have been handed over to logicians, who have countenanced the doctrine that their one function is to be judged as to their truth or falsehood.

The fact that propositions were first considered in connection with logic, and the moralistic preference for true propositions, have obscured the rôle of propositions in the actual world. Logicians only discuss the judgment of propositions. Indeed some philosophers fail to distinguish propositions from judgments; and most logicians consider propositions as merely appanages to judgments. Indeed Bradley does not mention 'propositions' in his *Logic*. He writes only of 'judgments.' Other authors define propositions as a component in judgment. The result is that false propositions have fared badly, thrown into the dust-heap, neglected. But in the real world it is more important that a proposition be interesting than that it be true. The importance of truth is, that it adds to interest.

The doctrine here laid down is that, in the realization of propositions, 'judgment' is a very rare component, and so is 'consciousness.' The existence of imaginative literature should have warned logicians that their narrow doctrine is absurd. It is difficult to believe that all logicians as they read Hamlet's speech, "To be, or not to be: . . ." commence by judging whether the initial proposition be true or false, and keep up the task of judgment throughout the whole thirty-five lines. Surely, at some point in the reading, judgment is eclipsed by aesthetic delight. The speech, for the theatre audience, is purely theoretical, a mere lure for feeling.

Again, consider strong religious emotion—consider a Christian meditating on the sayings in the Gospels. He is not judging 'true or false'; he is eliciting their value as elements in feeling. In fact, he may ground his judgment of truth upon his realization of value. But such a procedure is impossible, if the primary function of propositions is to be elements in judgments.

It is an essential doctrine in the philosophy of organism that the primary function of a proposition is to be relevant as a lure for feeling. For example, some propositions are the data of feelings with subjective forms such as to constitute those feelings to be the enjoyment of a joke. Other propositions are felt with feelings whose subjective forms are horror, disgust, or indignation. The ordinary logical account of 'propositions' expresses only a restricted aspect of their rôle in the universe, namely, when they are the data of feelings whose subjective forms are those of judgments.

To summarize this discussion of the general nature of a proposition: A proposition shares with an eternal object the character of indeterminateness, in that both are definite potentialities *for* actuality with undetermined realization *in* actuality. But they differ in that an eternal object refers to actuality with absolute generality, whereas a proposition refers to indicated logical subjects. Truth and falsehood always require some element of sheer givenness. Eternal objects cannot demonstrate what they are except in some given fact. The logical subjects of a proposition supply the element of givenness requisite for truth and falsehood.

3. *Physical Purposes versus Propositional Feelings*

An explicit discussion of the difference between physical purposes and propositional feelings will provide an introduction to the fourth phase of concrescence. Both physical purposes and propositional feelings are represented by circle c in Figure 2. The difference is that with a physical purpose the eternal object, which had been pried out of immanence into transcendence at the second phase of concrescence, sinks back into immanence at the third phase; the indetermination as to its ingressions, which had characterized it in the conceptual feeling, leaves it at the stage of physical purposes. The result is that it ceases to be a lure for feeling, and the process that is the concrescence of its subject comes to a halt with this "blank evaluation."

With a propositional feeling, on the other hand, the predicative pattern of the datum proposition does not lose its status as transcendent, but has its character as possibility enhanced. Whitehead states that the reason this happens is that in the datum proposition the component physical feeling has been reduced to the status of a bare logical subject. This prevents the predicative pattern from slipping back into immanence—there is not enough of a subject for it to merge with—and results in establishing the proposition as a lure for further integrations of feeling in Phase IV.

Whitehead is at no point excessively clear on these matters, but the following brief passages serve to corroborate this interpretation.

'Conceptual prehensions' can be 'pure' or 'impure.' An 'impure' prehension arises from the integration of a 'pure' conceptual prehension with a physical prehension originating in the physical pole. The datum of a pure conceptual prehension is an eternal object; the datum of an impure prehension is a proposition.[8]

The integration of a conceptual and physical prehension need not issue in an impure prehension [i.e., a propositional feeling]: the eternal object as a mere potentiality, undetermined as to its physical realization, may lose its indetermination, i.e. its universality, by integration with itself as an element in the realized definiteness of the physical datum of the physical prehension. In this case we obtain what is termed a 'physical purpose.'

In the more primitive type of comparative feelings [i.e., in physical purposes] indetermination as to its own ingressions is the aspect of the eternal object which is pushed into the background. In such a type of physical purposes the integration of a physical feeling and a conceptual feeling does not involve the reduction of the objective datum of the physical feeling to a multiplicity of bare logical subjects. The objective datum remains the nexus that it is, exemplifying the eternal objects whose ingression constitutes its definiteness. Also the indeterminateness as to its own ingressions is eliminated from the eternal object which is the datum of the conceptual feeling.

But with the growth of intensity in the mental pole, evidenced by the flash of novelty in appetition, the appetition takes the form of a 'propositional prehension.' In a propositional feeling the logical subjects have preserved their indicated particularity, but have lost their own real modes of objectification. A propositional feeling is a lure to creative emergence in the transcendent future. When it is functioning as a lure, the propositional feeling about the logical subjects of the proposition may in some subsequent phase promote decision involving intensification of some physical feeling of those subjects in the nexus.

---

[8] It is clear that "impure prehension" is another name for a "propositional feeling."

The propositions are lures for feelings, and give to feelings a definiteness of enjoyment and purpose which is absent in the blank evaluation of physical feeling into physical purpose. In this blank evaluation we have merely the determination of the comparative creative efficacies of the component feelings of actual entities. In a propositional feeling there is the 'hold up'—or, in its original sense, the epoch—of the valuation of the predicative pattern in its relevance to the definite logical subjects which are otherwise felt as definite elements in experience. There is the arrest of the emotional pattern round this sheer fact as a possibility, with the corresponding gain in distinctness of its relevance to the future. The particular possibility for the transcendent creativity—in the sense of its advance from subject to subject—this particular possibility has been picked out, held up, and clothed with emotion.

V

Phase IV—Complex Comparative Feelings

Complex comparative feelings, also termed intellectual feelings, are represented by circle d *in Figure 2. The datum for such a feeling, represented by bracket* z, *is the contrast between a nexus of actual entities (indicated by the lower leg of bracket* z) *and a proposition (indicated by the upper leg of* z) *the logical subjects of which are members of the nexus.*

In an intellectual feeling the datum is the generic contrast between a nexus of actual entities and a proposition with its logical subjects members of the nexus. It is the contrast between the affirmation of objectified fact in the physical feeling, and the mere potentiality, which is the negation of such affirmation, in the propositional feeling. It is the contrast between '*in fact*' and '*might be,*' in respect to particular instances in *this* actual world. This contrast is what has been termed the 'affirmation-negation contrast.'

The subjective form of the feeling of this contrast is consciousness. Thus in experience, consciousness arises by reason of intellectual feelings, and in proportion to the variety and intensity of such feelings. This account agrees with the plain facts of our conscious experience. Consciousness flickers; and even at its brightest, there is a small focal region of clear illumination, and a large penumbral region of experience which tells of intense experience in dim apprehension. The simplicity of clear consciousness is no measure of the complexity of complete experience. Also this character of our experience suggests that consciousness is the crown of experience, only occasionally attained, not its necessary base.

Consciousness is how we feel the affirmation-negation contrast. Conceptual feeling is the feeling of an unqualified negation; that is to say, it is the feeling of a definite eternal object with the definite extrusion of any particular realization. Consciousness requires that the objective datum should involve (as one side of a contrast) a qualified negative determined to some definite situation. This doctrine implies that there is no consciousness apart from propositions as one element in the objective datum. [But] consciousness requires more than the mere entertainment of theory. It is the feeling of the contrast of theory, as *mere* theory, with fact, as *mere* fact. This contrast holds whether or no the theory be correct.

The principle that I am adopting is that consciousness presupposes experience, and not experience consciousness. It is a special element in the subjective forms of some feelings. Thus an actual entity may, or may not, be conscious of some part of its experience. Its experience is its complete formal constitution, including its consciousness, if any.

Conceptual feelings do not necessarily involve consciousness; though there can be no conscious feelings which do not involve conceptual feelings as elements in the synthesis. Consciousness originates in the higher phases of integration and illuminates those phases with the greater clarity and distinctness. The consciousness belonging to an actual occasion is its sub-phase of intellectual supplementation, when that sub-phase is not purely trivial. This sub-phase is the eliciting, into feeling, of the full contrast between mere propositional potentiality and realized fact.

The important thing to grasp here is Whitehead's analysis of consciousness. Later on, in Chapter Six, he will use the crucial point that consciousness presupposes experience, and not experience consciousness, in his refutation of both Hume and Kant.

Only enough about complex comparative feelings has been

presented here to enable the account of consciousness to emerge clearly. Actually, Whitehead analyzes them carefully in what are some of the most difficult, but most suggestive, pages of PR. *(See Chapters IV and V of Part III. The editor has provided a guide to these chapters in his* A Whiteheadian Aesthetic, *Yale University Press, 1961, pp. 55–69.)*

This chapter has given the genetic analysis of actual occasions; it has traced the growth from phase to phase that is the process of concrescence. This growth is a growth of unity that culminates in the satisfaction. The genetic analysis of an actual entity discloses an initial stage of many feelings progressing through more and more complex stages of integration toward a terminal unity of feeling, toward one complex, fully determinate feeling termed the satisfaction of that actual entity.

VI

Satisfaction

The final phase in the process of concrescence, constituting an actual entity, is one complex, fully determinate feeling. This final phase is termed the 'satisfaction.' It is fully determinate (a) as to its genesis, (b) as to its objective character for the transcendent creativity, and (c) as to its prehension—positive or negative—of every item in its universe.

The subject completes itself during the process of concrescence by a self-criticism of its own incomplete phases. By this concrescence the multifold datum of the primary phase is gathered into the unity of the final satisfaction of feeling. The problem which the concrescence solves is, how the many components of the objective content are to be unified in one felt content with its complex subjective form. This one felt content is the 'satisfaction,' whereby the actual entity is its particular individual self, to use Descartes' phrase, 'requiring nothing but itself in order to exist.'

The process of the concrescence is a progressive integration of feelings controlled by their subjective forms. In this synthesis, feelings of an earlier phase sink into the components of some more complex feeling of a later phase. Thus each phase adds its element of novelty, until the final phase in which the one complex 'satisfaction' is reached. The analysis of the formal constitution of an actual entity has given three stages in the process of feeling: (i) the responsive phase, (ii) the supplemental stage,[9] and (iii) the satisfaction. The satisfaction is merely the culmination marking the evaporation of all indetermination; so that, in respect to all modes of feeling and to all entities in the universe, the satisfied actual entity embodies a determinate attitude of 'yes' or 'no.' Thus the satisfaction is the attainment of the private ideal which is the final cause of the concrescence. But the process itself lies in the two former phases.

The attainment of a peculiar definiteness is the final cause which animates a particular process; and its attainment halts its process. The term 'satisfaction' means the one complex fully determinate feeling which is the completed phase in the process.

'Satisfaction' is a generic term: there are specific differences between the 'satisfactions' of different entities, including gradations of intensity. These specific differences can only be expressed by the analysis of the components in the concrescence out of which the actual entity arises. The intensity of satisfaction is promoted by the 'order' in the phases from which concrescence arises and through which it passes; it is enfeebled by the 'disorder.' The components in the concrescence are thus 'values' contributory to the 'satisfaction.' The concrescence is thus the building up of a determinate 'satisfaction,' which constitutes the completion of the actual togetherness of the discrete components. The process of concrescence terminates with the attainment of a fully *determinate* 'satisfaction'; and the creativity thereby passes over into the 'given' primary phase for the concrescence of other actual entities. This transcendence is thereby established when there is attainment of *determinate* 'satisfaction' completing the antecedent entity. Completion is the perishing of immediacy: "It never really is."

In the conception of the actual entity in its phase of satisfaction, the entity has attained its individual separation from other things; it has absorbed the datum, and it has not yet lost itself in the swing back whereby its appetition becomes an element in the data of

[9] Note that this includes Phases II, III, and IV of Figure 2.

other entities superseding it. Time has stood still—if only it could. In other words, the 'satisfaction' of an entity can only be discussed in terms of the usefulness of that entity. It is a qualification of creativity. The tone of feeling embodied in this satisfaction passes into the world beyond, by reason of these objectifications. The world is self-creative; and the actual entity as self-creating creature passes into its immortal function of part-creator of the transcendent world. In its self-creation the actual entity is guided by its ideal of itself as individual satisfaction and as transcendent creator.

Chapter Four

NEXŪS AND THE
MACROCOSMIC

The first three chapters have presented Whitehead's account of actual entities, the building blocks of the universe. But actual entities are microcosmic, whereas we live in a macrocosmic world of trees, automobiles, houses, and people. Chapters Four and Five are concerned with the macrocosmic realm.

The present chapter concerns itself, first, with the bridge, namely transmutation, that enables Whitehead to move from the first realm to the second. Second, it analyzes the way actual entities group themselves into the aggregates, termed nexūs and societies, which are the macrocosmic entities. In this section the different types and levels of social organization are delineated. Third, the chapter utilizes the distinctions of Section II to "conjecture some fundamental principles of Psychological Physiology," which means, in effect, that these distinctions are applied to the traditional mind-body problem.

I

Transmutation

In Section III of Chapter Three, it was noted that Categoreal Obligation VI was being omitted for the moment. Categoreal Obligation VI is that of transmutation and will be considered now.

Transmutation is the operation whereby an aggregate of many actual occasions, forming a nexus, is prehended not as an aggregate, not as a many, but as a unity, as one macrocosmic entity.

The present section considers analogous feelings with diverse subjects 'scattered' throughout members of a nexus. It considers a single subject, subsequent to the nexus, prehending this multiplicity of scattered feelings as the *data* for a corresponding multiplicity of its own simple physical feelings, some pure and some hybrid. It then formulates the process by which in that subject an analogy between these various feelings—constituted by one eternal object, of whatever complexity, implicated in the various analogous *data* of these feelings—is, by a supervening process of integration, converted into one feeling having for its datum the specific contrast between the nexus as one entity and that eternal object.

CATEGORY VI. THE CATEGORY OF TRANSMUTATION. When (in accordance with Category IV, or with Categories IV and V) one, and the same, conceptual feeling is derived impartially by a prehending subject from its analogous simple physical feelings of various actual entities, then in a subsequent phase of integration— of these simple physical feelings together with the derivate conceptual feeling—the prehending subject may transmute the *datum* of this conceptual feeling into a characteristic of some *nexus* containing those prehended actual entities among its members, or of some part of that nexus. In this way the nexus (or its part), thus characterized, is the objective datum of a feeling entertained by this prehending subject.

Such a transmutation of simple physical feelings of many actualities into one physical feeling of a nexus as one, is called a 'transmuted feeling.' It is evident that the complete datum of the transmuted feeling is a contrast, namely, 'the nexus, as one, in contrast with the eternal object.' This type of contrast is one of the meanings of the notion 'qualification of physical substance by quality.'

Figure 3 will aid in understanding transmutation. The datum occasions at the left of Figure 3 are all characterized by, say, the eternal object "red." A series of simple physical feelings, a, a_1, a_2, and so on, occurs at Phase I of the concrescing subject, each

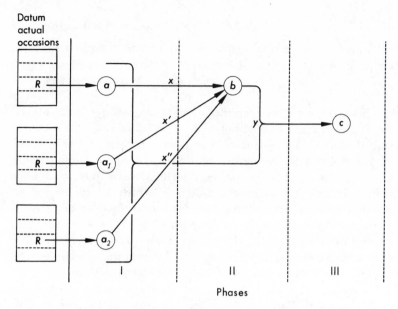

Figure 3. Transmutation.

item of which objectifies its datum from a perspective including
the eternal object "red." At Phase II a single conceptual feeling
emerges, represented by circle b, which has the eternal object "red"
as datum. The conceptual feeling is derived from all the a's and
hence has impartial relevance to the whole series of a's. The
simple comparative feeling c, which arises in the third phase, contrasts (as indicated by bracket y) b with all the involved physical
feelings. The feeling c prehends the entire nexus of datum occasions
as one entity qualified by the eternal object "red"—c is a transmuted physical feeling aware of one macrocosmic entity where before there were many microcosmic entities.

In the paragraphs that follow, Whitehead explicates the concept
of transmutation.

This category is the way in which the philosophy of organism,
which is an atomic theory of actuality, meets a perplexity which
is inherent in all monadic cosmologies. Leibniz in his *Monadology*
meets the same difficulty by a theory of 'confused' perception. But
he fails to make clear how 'confusion' originates.

In order to understand this categoreal condition, it must be noted that the integration of simple physical feelings into a complex physical feeling only provides for the various actual entities of the nexus being felt as separate entities requiring each other. We have to account for the substitution of the one nexus in place of its component actual entities. This is Leibniz's problem which arises in his *Monadology*. He solves the problem by an unanalysed doctrine of 'confusion.' Some category is required to provide a physical feeling of a nexus as one entity with its own categoreal type of existence. This one physical feeling in the final subject is derived by transmutation from the various analogous physical feelings entertained by the various members of the nexus.

In the prehending subject, these analogous, pure physical feelings originate a conceptual feeling. This conceptual feeling has an impartial relevance to the various simple physical feelings of the various members of the nexus. It is this impartiality of the conceptual feeling which leads to the integration in which the many members of the nexus are collected into the one nexus which they form, and in which that nexus is set in contrast to the one eternal object which has emerged from their analogies.

But there may be reversions to be considered, that is to say, reversions with impartial reference throughout the nexus. The reversion may originate in the separate actualities of the nexus, or in the final prehending subject, or there may be a double reversion involving both sources. Thus we must allow for the possibility of diverse reverted feelings, each with impartial reference. In so far as there is concordance and the reversions are dominant, there will issue one conceptual feeling of enhanced intensity. When there is discordance among these various conceptual feelings, there will be elimination, and in general no transmutation. But when, from some (or all) of these sources of impartial conceptual feelings, one dominant impartial conceptual feeling emerges with adequate intensity, transmutation will supervene.

This impartiality of reference has then been transmuted into the physical feeling of that nexus, whole or partial, contrasted with some one eternal object. It will be noted that this one impartial conceptual feeling is an essential element of the process, whereby an impartial reference to the whole nexus is introduced. Otherwise there would be no element to transmute particular relevancies to the many members into general relevance to the whole.

The eternal object which characterizes the nexus in this physical feeling may be an eternal object characterizing the analogous physical feelings, belonging to all, or some, of the members of the nexus. In this case, the nexus as a whole derives a character which in some way belongs to its various members.

Also the eternal object may be the datum of a reverted conceptual feeling, only indirectly derived from the members of the original nexus. In this case, the transmuted feeling of the nexus introduces novelty; and in unfortunate cases this novelty may be termed 'error.' New forms enter into positive realizations first as conceptual experience, and are then transmuted into physical experience. But all the same, the transmuted feeling, whatever be its history of transmutation, is a definite physical fact whereby the final subject prehends the nexus. For example, colour-blindness may be called 'error'; but nevertheless, it is a physical fact. A transmuted feeling comes under the definition of a physical feeling. In a simple physical feeling, the initial datum is a single actual entity; in a transmuted feeling, the objective datum is a nexus of actual entities. Simple physical feelings and transmuted feelings make up the class of physical feelings.

Our usual way of consciously prehending the world is by these transmuted physical feelings. The intensity arising from the force of repetition makes this transmuted perception to be the prominent type of those feelings which in further integrations acquire consciousness as an element in their subjective forms. It represents a simplification of physical feeling, effected in the course of integration. It is only when we are consciously aware of alien mentalities that we even approximate to the conscious prehension of a single actual entity.

It is evident that adversion and aversion, and also the Category of Transmutation, only have importance in the case of high-grade organisms. They constitute the first step towards intellectual mentality, though in themselves they do not amount to consciousness. But an actual entity which includes these operations must have an important intensity of conceptual feelings able to mask and fuse the simple physical feelings.

The examination of the Category of Transmutation shows that the approach to intellectuality consists in the gain of a power of abstraction. The irrelevant multiplicity of detail is eliminated, and emphasis is laid on the elements of systematic order in the actual world. Apart from transmutation our feeble intellectual operations would fail to penetrate into the dominant characteristics of things. We can only understand by discarding. The low-grade organism is merely the summation of the forms of energy which flow in upon it in all their multiplicity of detail. It receives, and it transmits; but it fails to simplify into intelligible system.

Transmutation is the way in which the actual world is felt as a community, and is so felt in virtue of its prevalent order. For it arises by reason of the analogies between the various members of the prehended nexus, and eliminates their differences. In so far as there is trivial order, there must be trivialized actual entities. The right co-ordination of the negative prehensions is one secret of mental progress; but unless some systematic scheme of related-ness characterizes the environment, there will be nothing left whereby to constitute vivid prehension of the world.

II

Nexūs and Order

The last paragraph makes it clear that transmutation presup-poses order. A nexus is a "systematic scheme of relatedness"; nexūs provide the "prevalent order" that enables the actual world to be "felt as a community" via transmutation. In this section Whitehead discusses nexūs, and the special sorts of nexūs called societies, which are presupposed by transmutation.

Actual entities involve each other by reason of their prehensions of each other. There are thus real individual facts of the together-ness of actual entities, which are real, individual, and particular, in the same sense in which actual entities and the prehensions are real, individual, and particular. Any such particular fact of to-getherness among actual entities is called a 'nexus' (plural form is written 'nexūs'). A nexus is a set of actual entities in the unity of the relatedness constituted by their prehensions of each other, or —what is the same thing conversely expressed—constituted by their objectifications in each other. For some purposes a nexus of many actualities can be treated as though it were one actuality. This is what we habitually do in the case of the span of life of a molecule, or of a piece of rock, or of a human body.

A nexus is normally four-dimensional. A tree is a complex nexus. At any given moment it is composed of a generation of actual occasions spatially related in a three-dimensional pattern. But the total nexus that is the tree is temporally thick also—it consists of generation after generation of actual occasions succeeding one another.

One limiting type of nexus plays an important role later on in the chapter. A purely temporal nexus nearly lacks the dimension of space; it includes no pair of contemporary occasions and hence is "a mere thread of temporal transition from occasion to occasion." Such a nexus is termed a personally ordered nexus and constitutes the only kind of society of occasions properly referred to as a person. This notion is essential to Whitehead's resolution of the mind-body problem in Section III.

A 'society,' in the sense in which that term is here used, is a nexus with social order. A nexus enjoys 'social order' where (i) there is a common element of form illustrated in the definiteness of each of its included actual entities, and (ii) this common element of form arises in each member of the nexus by reason of the conditions imposed upon it by its prehensions of some other members of the nexus, and (iii) these prehensions impose that condition of reproduction by reason of their inclusion of positive feelings of that common form. Such a nexus is called a 'society,' and the common form is the 'defining characteristic' of the society. The notion of 'defining characteristic' is allied to the Aristotelian notion 'substantial form.'

The common element of form is simply a complex eternal object exemplified in each member of the nexus. But the social order of the nexus is not the mere fact of this common form exhibited by all its members. The point of a 'society,' as the term is here used, is that it is self-sustaining; in other words, that it is its own reason. Thus a society is more than a set of entities to which the same class-name applies: that is to say, it involves more than a merely mathematical conception of 'order.' To constitute a society, the class-name has got to apply to each member, by reason of genetic derivation from other members of that same society. The members of the society are alike because, by reason of their common character, they impose on other members of the society the conditions which lead to that likeness.

For example, the life of man is an historic route of actual occasions which in a marked degree—to be discussed more fully later—inherit from each other. That set of occasions, dating from his first acquirement of the Greek language and including all those occasions up to his loss of any adequate knowledge of that language, constitutes a society in reference to knowledge of the Greek language. Such knowledge is a common characteristic inherited from occasion to occasion along the historic route.

Thus a society is, for each of its members, an environment with some element of order in it, persisting by reason of the genetic relations between its own members. Such an element of order is the order prevalent in the society.

But there is no society in isolation. Every society must be considered with its background of a wider environment of actual entities, which also contribute their objectifications to which the members of the society must conform. Thus the given contributions of the environment must at least be permissive of the self-sustenance of the society. Also, in proportion to its importance, this background must contribute those general characters which the more special character of the society presupposes for its members. But this means that the environment, together with the society in question, must form a larger society in respect to some more general characters than those defining the society from which we started. Thus we arrive at the principle that every society requires a social background, of which it is itself a part. In reference to any given society the world of actual entities is to be conceived as forming a background in layers of social order, the defining characteristics becoming wider and more general as we widen the background.

The causal laws which dominate a social environment are the

product of the defining characteristic of that society. But the society is only efficient through its individual members. Thus in a society, the members can only exist by reason of the laws which dominate the society, and the laws only come into being by reason of the analogous characters of the members of the society.

*   *   *

The remainder of this chapter will be devoted to a discussion—largely conjectural—of the hierarchy of societies composing our present epoch. In this way, the preceding discussion of 'order' may be elucidated. It is to be carefully noted that we are now deserting metaphysical generality. We shall be considering the more special possibilities of explanation consistent with our general cosmological doctrine, but not necessitated by it.

The metaphysical characteristics of an actual entity are those that apply to all actual entities. In the previous chapters Whitehead's account of the general characteristics shared by all actual entities has been presented in detail. The investigation now turns to the particular ways in which certain limited groups of actual entities are joined in social relationships. In this sense metaphysical generality is being deserted.

Whitehead's account of the "hierarchy of societies" can best be visualized under the image of a set of Chinese boxes, each box lodged within a larger box. The outermost box is labeled "the society of pure extension"—this constitutes the most general form of social relatedness presently conceivable. Within this "box" a more specialized social order is distinguishable, that of a geometrical society. Within the geometrical society can be found the even more specific set of social relations constitutive of the electromagnetic society that forms our own cosmic epoch. The narrower, more specialized societies do not repudiate the order of the wider societies within which they occur; rather, they presuppose those wider social orderings and give to them additional, more specialized and limited types of social order.

The Chinese-boxes image breaks down in that there may be many boxes, each with its sub-societies, within any given box. For example, within a given geometrical society there may be an electro-

magnetic society like the one in which we live plus another society, physically removed from it, made up of "antimatter," of matter in which the electrical characteristics are the reverse of what we know. Recent experiments indicate that such a notion is not mere fantasy. Also, within the society of pure extension various geometrical societies are possible. Our own four-dimensional system may be paralleled by seven-, eight-, or nine-dimensional systems, and within our particular four-dimensional system there may be various societies differentiated by more specialized geometrical relationships in virtue of which straight lines are defined.

The physical world is bound together by a general type of relatedness which constitutes it into an extensive continuum. In these general properties of extensive connection, we discern the defining characteristic of a vast nexus extending far beyond our immediate cosmic epoch. It contains in itself other epochs, with more particular characteristics incompatible with each other. This ultimate, vast society constitutes the whole environment within which our epoch is set, so far as systematic characteristics are discernible by us in our present stage of development.

Our logical analysis, in company with immediate intuition (*inspectio*), enables us to discern a more special society within the society of pure extension. This is the 'geometrical' society. In this society those specialized relationships hold, in virtue of which straight lines are defined. Systematic geometry is illustrated in such a geometrical society; and metrical relationships can be defined in terms of the analogies of function within the scheme of any one systematic geometry.

Our present cosmic epoch is formed by an 'electromagnetic' society, which is a more special society contained within the geometric society. In this society yet more special defining characteristics obtain. These characteristics presuppose those of the two wider societies within which the 'electromagnetic' society is contained.

The electromagnetic society exhibits the physical electromagnetic field which is the topic of physical science. The members of this nexus are the electromagnetic occasions. Thus our present epoch is dominated by a society of electromagnetic occasions. In so far as this dominance approaches completeness, the systematic law which physics seeks is absolutely dominant. In so far as the dominance is incomplete, the obedience is a statistical fact with its corresponding lapses.

Thus the physical relations, the geometrical relations of measurement, the dimensional relations, and the various grades of extensive relations, involved in the physical and geometrical theory of nature, are derivative from a series of societies of increasing width of prevalence, the more special societies being included in the wider societies. This situation constitutes the physical and geometrical order of nature. Beyond these societies there is disorder, where 'disorder' is a relative term expressing the lack of importance possessed by the defining characteristics of the societies in question beyond their own bounds. When those societies decay, it will not mean that their defining characteristics cease to exist; but that they lapse into unimportance for the actual entities in question. The term 'disorder' refers to a society only partially influential in impressing its characteristics in the form of prevalent laws. This doctrine, that order is a social product, appears in modern science as the statistical theory of the laws of nature, and in the emphasis on genetic relation.

There is not any perfect attainment of an ideal order whereby the indefinite endurance of a society is secured. A society arises from disorder, where 'disorder' is defined by reference to the ideal for that society; the favourable background of a larger environment either itself decays, or ceases to favour the persistence of the society after some stage of growth: the society then ceases to reproduce its members, and finally after a stage of decay passes out of existence. Thus a system of 'laws' determining reproduction in some portion of the universe gradually rises into dominance; it has its stage of endurance, and passes out of existence with the decay of the society from which it emanates.

The arbitrary, as it were 'given,' elements in the laws of nature warn us that we are in a special cosmic epoch. Here the phrase 'cosmic epoch' is used to mean that widest society of actual entities whose immediate relevance to ourselves is traceable. This epoch is characterized by electronic and protonic actual entities, and by yet more ultimate actual entities which can be dimly discerned in the quanta of energy. Maxwell's equations of the electromagnetic field hold sway by reason of the throngs of electrons and of protons. Also each electron is a society of electronic occasions, and each proton is a society of protonic occasions. These occasions are the reasons for the electromagnetic laws; but their capacity for reproduction, whereby each electron and each proton has a long life, and whereby new electrons and new protons come into being, is itself due to these same laws. But there is disorder in the sense that the laws are not perfectly obeyed, and that the reproduction is mingled with instances of failure. The immanence of God gives reason for the belief that pure chaos is intrinsically impossible. At the other end of the scale, the immensity of the world negatives the belief that any state of order can be so established that beyond it there can be no progress. This belief in a final order, popular in religious and philosophic thought, seems to be due to the prevalent fallacy that all types of seriality necessarily involve terminal instances. It follows that Tennyson's phrase,

> . . . one far-off divine event
> To which the whole creation moves,

presents a fallacious conception of the universe.

So far Whitehead has specified the broad, general levels of social order within which our cosmic epoch is set. He turns now to a consideration of the much more specialized types of social order to be observed structuring the world about us.

An electron or a proton is a society of electronic or protonic occasions. More specialized forms of social order incorporate electrons and protons into atoms, atoms into molecules, molecules into living cells, and cells into vegetable and animal bodies. Atoms, molecules, cells, and bodies are complex societies, hierarchies of societies within societies. Such complex societies are termed "structured" societies.

But in its turn, this electromagnetic society would provide no adequate order for the production of individual occasions realizing peculiar 'intensities' of experience unless it were pervaded by more special societies, vehicles of such order. The physical world exhibits a bewildering complexity of such societies, favouring each other, competing with each other. The most general examples of such societies are the regular trains of waves, individual electrons, protons, individual molecules, societies of molecules such as in- organic bodies, living cells, and societies of cells such as vegetable and animal bodies.

The notion of a society which includes subordinate societies and nexūs with a definite pattern of structural inter-relations, must be introduced. Such societies will be termed 'structured.'

A structured society as a whole provides a favourable environ- ment for the subordinate societies which it harbours within itself. Also the whole society must be set in a wider environment per- missive of its continuance. Some of the component groups of occa- sions in a structured society can be termed 'subordinate societies.' But other such groups must be given the wider designation of 'sub- ordinate nexūs.' The distinction arises because in some instances a group of occasions, such as, for example, a particular enduring entity, could have retained the dominant features of its defining charac- teristic in the general environment, apart from the structured soci- ety. It would have lost some features; in other words, the analogous sort of enduring entity in the general environment is, in its mode of definiteness, not quite identical with the enduring entity within the structured environment. But, abstracting such additional details from the generalized defining characteristic, the enduring object with that generalized characteristic may be conceived as independ- ent of the structured society within which it finds itself. For example, we speak of a molecule within a living cell, because its general molecular features are independent of the environment of the cell. Thus a molecule is a subordinate society in the structured society which we call the 'living cell.'

But there may be other nexūs included in a structured society which, excepting the general systematic characteristics of the external environment, present no features capable of genetically sustaining themselves apart from the special environment provided by that structured society. It is misleading, therefore, to term such a [subordinate] nexus a 'society' when it is being considered in ab- straction from the whole structured society. In such an ab- straction it can be assigned no 'social' features. Recurring to the example of a living cell, it will be argued that the occasions com- posing the 'empty' space within the cell exhibit special features which analogous occasions outside the cell are devoid of. Thus the nexus, which is the empty space within a living cell, is called a 'subordinate nexus,' but not a 'subordinate society.'

Molecules are structured societies, and so in all probability are separate electrons and protons. Crystals are structured societies. But gases are not structured societies in any important sense of that term; although their individual molecules are structured societies.

It must be remembered that each individual occasion within a special form of society includes features which do not occur in analogous occasions in the external environment. The first stage of systematic investigation must always be the identification of analogies between occasions within the society and occasions without it. The second stage is constituted by the more subtle procedure of noting the differences between behaviour within and without the society, differences of behaviour exhibited by occasions which also have close analogies to each other. The history of science is marked by the vehement, dogmatic denial of such differences, until they are found out. An obvious instance of such distinction of behaviour is afforded by the notion of the deformation of the shape of an electron according to variations in its physical situation.

A 'structured society' may be more or less 'complex' in respect to the multiplicity of its associated sub-societies and sub-nexūs and to the intricacy of their structural pattern. An ordinary physical object, which has temporal endurance, is a society. In the ideally simple case, it has personal order.

A nexus enjoys 'personal order' when (a) it is a 'society,' and (β) when the genetic relatedness of its members orders these members 'serially.' Thus the nexus forms a single line of inheritance of its defining characteristic. Such a nexus is called an 'enduring object.' It might have been termed a 'person,' in the legal sense of that term. But unfortunately 'person' suggests the notion of consciousness, so that its use would lead to misunderstanding. The nexus 'sustains a character,' and this is one of the meanings of the Latin word *persona*. But an 'enduring object,' *qua* 'person,' does more than sustain a character. For this sustenance arises out of the special genetic relations among the members of the nexus.

A society may (or may not) be analysable into many strands of 'enduring objects.' This will be the case for most ordinary physical objects. These enduring objects and 'societies,' analysable into strands of enduring objects, are the permanent entities which enjoy adventures of change throughout time and space. For example, they form the subject-matter of the science of dynamics. Actual entities perish, but do not change; they are what they are. A nexus which (i) enjoys social order, and (ii) is analysable into strands of enduring objects may be termed a 'corpuscular society.' A society may be more or less corpuscular, according to the relative importance of the defining characteristics of the various enduring objects compared to that of the defining characteristic of the whole corpuscular nexus.[1]

A structured society which is highly complex can be correspondingly favourable to intensity of satisfaction for certain sets of its component members. This intensity arises by reason of the ordered complexity of the contrasts which the society stages for these components. The structural relations gather intensity from this intensity in the individual experiences. Thus the growth of a complex structured society exemplifies the general purpose pervading nature. The mere complexity of givenness which procures incompatibilities has been superseded by the complexity of order which procures contrasts.

The doctrine that every society requires a wider social environment leads to the distinction that a society may be more or less 'stabilized' in reference to certain sorts of changes in that environment. A society is 'stabilized' in reference to a species of change when it can persist through an environment whose relevant parts exhibit that sort of change. If the society would cease to persist through an environment with that sort of heterogeneity, then the society is in that respect 'unstable.' A complex society which is stable provided that the environment exhibits certain features, is said to be 'specialized' in respect to those features. The notion of 'specialization' seems to include both that of 'complexity' and that of strictly conditioned 'stability.'

An unspecialized society can survive through important changes

[1] A clear example of this distinction occurs in American history. The colonies under the Articles of Confederation were a more corpuscular society than were the colonies under the Constitution because under the Articles the characteristics of the component states were more important relative to the characteristics of the Central Government than they were under the Constitution.

in its environment. This means that it can take on different functions in respect to its relationship to a changing environment. In general the defining characteristic of such a society will not include any particular determination of structural pattern. By reason of this flexibility of structural pattern, the society can adopt that special pattern adapted to the circumstances of the moment. Thus an unspecialized society is apt to be deficient in structural pattern, when viewed as a whole.

Thus in general an unspecialized society does not secure conditions favourable for intensity of satisfaction among its members. Whereas a structured society with a high grade of complexity will in general be deficient in survival value. In other words, such societies will in general be 'specialized' in the sense of requiring a very special sort of environment.

Thus the problem for Nature is the production of societies which are 'structured' with a high 'complexity,' and which are at the same time 'unspecialized.' In this way, intensity is mated with survival.

There are two ways in which structured societies have solved this problem. Both ways depend on that enhancement of the mental pole, which is a factor in intensity of experience. One way is by eliciting a massive average objectification of a nexus, while eliminating the detailed diversities of the various members of the nexus in question. This method, in fact, employs the device of blocking out unwelcome detail. It depends on the fundamental truth that objectification is abstraction. It utilizes this abstraction inherent in objectification so as to dismiss the thwarting elements of a nexus into negative prehensions. At the same time the complex intensity in the structured society is supported by the massive objectifications of the many environmental nexūs, each in its unity as *one* nexus, and not in its multiplicity as *many* actual occasions.

Material bodies belong to the lowest grade of structured societies which are obvious to our gross apprehensions. They comprise societies of various types of complexity—crystals, rocks, planets, and suns. Such bodies are easily the most long-lived of the structured societies known to us, capable of being traced through their individual life-histories.

The second way of solving the problem is by an initiative in conceptual prehensions, i.e., in appetition. The purpose of this initiative is to receive the novel elements of the environment into explicit feelings with such subjective forms as conciliate them with the complex experiences proper to members of the structured society. Thus in each concrescent occasion its subjective aim originates novelty to match the novelty of the environment.

In the case of the higher organisms, this conceptual initiative amounts to *thinking* about the diverse experiences; in the case of lower organisms this conceptual initiative merely amounts to thoughtless adjustment of aesthetic emphasis in obedience to an ideal of harmony. In either case the creative determination which transcends the occasion in question has been deflected by an impulse original to that occasion. This deflection in general originates a self-preservative reaction throughout the whole society. It may be unfortunate or inadequate; and in the case of persistent failure we are in the province of pathology.

Structured societies in which the second mode of solution has importance are termed 'living.' It is obvious that a structured society may have more or less 'life,' and that there is no absolute gap between 'living' and 'non-living' societies. For certain purposes, whatever 'life' there is in a society may be important; and for other purposes, unimportant.

A structured society in which the second mode is unimportant, and the first mode is important will be termed 'inorganic.'

In accordance with this doctrine of 'life,' the primary meaning of 'life' is the origination of conceptual novelty—novelty of appetition. Such origination can only occur in accordance with the Category of Reversion. Thus a society is only to be termed 'living' in a derivative sense. A 'living society' is one which includes some 'living occasions.' Thus a society may be more or less 'living,' according to the prevalence in it of living occasions. Also an occasion may be more or less living according to the relative importance of the novel factors in its final satisfaction.

There is yet another factor in 'living' societies which requires more detached analysis. A structured society consists in the patterned intertwining of various nexūs with markedly diverse defining characteristics. Some of these nexūs are of lower types than others, and some will be of markedly higher types. There will be the 'subservient' nexūs and the 'regnant' nexūs within the same structured society. This structured society will provide the immediate environment which sustains each of its sub-societies, subservient and regnant alike. In a living society only some of its nexūs will be such that the mental poles of all their members have any *original* reactions. These will be its 'entirely living' nexūs, and in practice a society is only called 'living' when such nexūs are regnant. Thus a living society involves nexūs which are 'inorganic,' and nexūs which are inorganic do not need the protection of the whole 'living' society for their survival in a changing external environment. Such nexūs are [subordinate] societies. But 'entirely living' nexūs do require such protection, if they are to survive.[2] A complex inorganic system of interaction is built up for the protection of the 'entirely living' nexūs, and the originative actions of the living elements are protective of the whole system. On the other hand, the reactions of the whole system provide the intimate environment required by the 'entirely living' nexūs. We do not know of any living society devoid of its subservient apparatus of inorganic societies. It must be remembered that an integral living society, as we know it, not only includes the subservient inorganic apparatus, but also includes many living nexūs—at least one for each 'cell.'

III
Psychological Physiology

In this section the categories of social organization previously presented are focused upon an analysis of higher forms of life with the purpose of providing a systematic resolution of the traditional problem of the relationship holding between mind and body. Such an investigation is what Whitehead would term "psychological physiology." See the entry Structured society *in the Glossary for a synoptic account of the relationships to be here established.*

'Physical Physiology' deals with the subservient inorganic apparatus; and 'Psychological Physiology' seeks to deal with 'entirely living' nexūs, partly in abstraction from the inorganic apparatus, and partly in respect to their response to the inorganic apparatus,

[2] Hence "entirely living" nexūs are subordinate nexūs and not subordinate societies.

and partly in regard to their response to each other. Physical Physiology has, in the last century, established itself as a unified science; Psychological Physiology is still in the process of incubation.

It will throw light upon the cosmology of the philosophy of organism to conjecture some fundamental principles of Psychological Physiology as suggested by that cosmology and by the preceding conjectures concerning the 'societies' of our epoch. These principles are not necessitated by this cosmology; but they seem to be the simplest principles which are both consonant with that cosmology, and also fit the facts.

In the first instance, consider a single living cell. Such a cell includes subservient inorganic societies, such as molecules and electrons. Thus, the cell is an 'animal body'; and we must presuppose the 'physical physiology' proper to this instance. But what of the individual living occasions?

The first question to be asked is as to whether the living occasions, in abstraction from the inorganic occasions of the animal body, form a corpuscular sub-society, so that each living occasion is a member of an enduring entity with its personal order. In particular we may ask whether this corpuscular society reduces to the extreme instance of such a society, namely, to one enduring entity with its one personal order.

The evidence before us is of course extremely slight; but so far as it goes, it suggests a negative answer to both these questions. A cell gives no evidence whatever of a single unified mentality, guided in each of its occasions by inheritance from its own past. The problem to be solved is that of a certain originality in the response of a cell to external stimulus. The theory of an enduring entity with its inherited mentality gives us a reason why this mentality should be swayed by its own past. We ask for something original at the moment, and we are provided with a reason for limiting originality. Life is a bid for freedom: an enduring entity binds any one of its occasions to the line of its ancestry. The doctrine of the enduring soul with its permanent characteristics is exactly the irrelevant answer to the problem which life presents. That problem is, How can there be originality? And the answer explains how the soul need be no more original than a stone.

The theory of a corpuscular society, made up of many enduring entities, fits the evidence no better. The same objections apply. The root fact is that 'endurance' is a device whereby an occasion is peculiarly bound by a single line of physical ancestry, while 'life' means novelty, introduced in accordance with the Category of Conceptual Reversion. There are the same objections to *many* traditions as there are to *one* tradition. What has to be explained is originality of response to stimulus. This amounts to the doctrine that an organism is 'alive' when in some measure its reactions are inexplicable by *any* tradition of pure physical inheritance.

Explanation by 'tradition' is merely another phraseology for explanation by 'efficient cause.' We require explanation by 'final cause.' Thus a single occasion is alive when the subjective aim which determines its process of concrescence has introduced a novelty of definiteness not to be found in the inherited data of its primary phase. The novelty is introduced conceptually and disturbs the inherited 'responsive' adjustment of subjective forms. It alters the 'values,' in the artist's sense of that term.

It follows from these considerations that in abstraction from its animal body an 'entirely living' nexus is not properly a society at all, since 'life' cannot be a defining characteristic. It is the name for originality, and not for tradition. The mere response to stimulus is characteristic of all societies whether inorganic or alive. Action and reaction are bound together. The characteristic of life is reaction adapted to the capture of intensity, under a large variety of circumstances. But the reaction is dictated by the present and not by the past. It is the clutch at vivid immediacy.

Another characteristic of a living society is that it requires food. In a museum the crystals are kept under glass cases; in zoological gardens the animals are fed. Having regard to the universality of reactions with environment, the distinction is not quite absolute. It cannot, however, be ignored. The crystals are not agencies requiring the destruction of elaborate societies derived from the environment; a living society is such an agency. The societies which it destroys are its food. This food is destroyed by dissolving it into somewhat simpler social elements. It has been robbed of something. Thus, all societies require interplay with their environment; and in the case of living societies this interplay takes the form of robbery. The living society may, or may not, be a higher type of organism than the food which it disintegrates. But whether or no it be for the general good, life is robbery. It is at this point that with life morals become acute. The robber requires justification.

The primordial appetitions which jointly constitute God's purpose are seeking intensity, and not preservation. Because they are primordial, there is nothing to preserve. He, in his primordial nature, is unmoved by love for this particular, or that particular; for in this foundational process of creativity, there are no preconstituted particulars. In the foundations of his being, God is indifferent alike to preservation and to novelty. He cares not whether an immediate occasion be old or new, so far as concerns derivation from its ancestry. His aim for it is depth of satisfaction as an intermediate step towards the fulfilment of his own being. His tenderness is directed toward each actual occasion, as it arises.

Thus God's purpose in the creative advance is the evocation of intensities. The evocation of societies is purely subsidiary to this absolute end. The characteristic of a living society is that a complex structure of inorganic societies is woven together for the production of a non-social nexus characterized by the intense physical experiences of its members. But such an experience is derivate from the complex order of the material animal body, and not from the simple 'personal order' of past occasions with analogous experience. There is intense experience without the shackle of reiteration from the past. This is the condition for spontaneity of conceptual reaction. The conclusion to be drawn from this argument is that life is a characteristic of 'empty space' and not of space 'occupied' by any corpuscular society. In a nexus of living occasions, there is a certain social deficiency. Life lurks in the interstices of each living cell, and in the interstices of the brain. In the history of a living society, its more vivid manifestations wander to whatever quarter is receiving from the animal body an enormous variety of physical experience. This experience, if treated inorganically, must be reduced to compatibility by the normal adjustments of mere responsive reception. This means the dismissal of incompatible elements into negative prehensions.

The complexity of the animal body is so ordered that in the critical portions of its interstices the varied datum of physical experience is complex, and on the edge of a compatibility beyond that to be achieved by mere inorganic treatment. A novel conceptual prehension disturbs the subjective forms of the initial responsive phase. Some negative prehensions are thus avoided, and higher contrasts are introduced into experience.

So far as the functioning of the animal body is concerned, the total result is that the transmission of physical influence, through the empty space within it, has not been entirely in conformity with the physical laws holding for inorganic societies. The molecules within an animal body exhibit certain peculiarities of behaviour not to be detected outside an animal body. In fact, living societies illustrate the doctrine that the laws of nature develop together with societies which constitute an epoch. There are statistical expressions of the prevalent types of interaction. In a living cell, the statistical balance has been disturbed.

The connection of 'food' with 'life' is now evident. The highly complex inorganic societies required for the structure of a cell, or other living body, lose their stability amid the diversity of the environment. But, in the physical field of empty space produced by the originality of living occasions, chemical dissociations and associations take place which would not otherwise occur. The structure is breaking down and being repaired. The food is that supply of highly complex societies from the outside which, under the influence of life, will enter into the necessary associations to repair the waste. Thus life acts as though it were a catalytic agent.

The short summary of this account of a living cell is as follows: (i) an extremely complex and delicately poised chemical structure; (ii) for the occasions in the interstitial 'empty' space a complex objective datum derived from this complex structure; (iii) under normal 'responsive' treatment, devoid of originality, the complex detail reduced to physical simplicity by negative prehensions; (iv) this detail preserved for positive feeling by the emotional and purposive readjustments produced by originality of conceptual feeling (appetition); (v) the physical distortion of the field, leading to instability of the structure; (vi) the structure accepting repair by food from the environment.

The complexity of nature is inexhaustible. So far we have argued that the nature of life is not to be sought by its identification with some society of occasions, which are living in virtue of the defining characteristic of that society. An 'entirely living' nexus is, in respect to its life, not social. Each member of the nexus derives the necessities of its being from its prehensions of its complex social environment; by itself the nexus lacks the genetic power which belongs to 'societies.' But a living nexus, though non-social in virtue of its 'life,' may support a thread of personal order along some historical route of its members. Such an enduring entity is a 'living person.' It is not of the essence of life to be a living person. Indeed a living person requires that its immediate environment be a living, non-social nexus.

The defining characteristic of a living person is some definite type of hybrid prehensions transmitted from occasion to occasion of its existence. A 'hybrid' prehension is the prehension by one subject of a conceptual prehension, or of an 'impure' prehension, belonging to the mentality of another subject. By this transmission the mental originality of the living occasions receives a character and a depth. In this way originality is both 'canalized'—to use Bergson's word— and intensified. Its range is widened within limits. Apart from canalization, depth of originality would spell disaster for the animal body. With it, personal mentality can be evolved, so as to combine its individual originality with the safety of the material organism on which it depends. Thus life turns back into society: it binds originality within bounds, and gains the massiveness due to re-iterated character.

In the case of single cells, of vegetation, and of the lower forms of animal life, we have no ground for conjecturing living personality. But in the case of the higher animals there is central direction, which suggests that in their case each animal body harbours a living person, or living persons. Our own self-consciousness is direct awareness of ourselves as such persons. There are limits to such unified control, which indicate dissociation of personality, multiple personalities in successive alternations, and even multiple personalities in joint possession. This last case belongs to the pathology of religion, and in primitive times has been interpreted as demoniac possession. Thus, though life in its essence is the gain of intensity through freedom, yet it can also submit to canalization and so gain the massiveness of order. But it is not necessary merely to presuppose the drastic case of personal order. We may conjecture, though without much evidence, that even in the lowest form of life the entirely living nexus is canalized into some faint form of mutual conformity. Such conformity amounts to social order depending on hybrid prehensions of originalities in the mental poles of the antecedent members of the nexus. The survival power, arising from adaptation and regeneration, is thus explained. Thus life is a passage from physical order to pure mental originality, and from pure mental originality to canalized mental originality. It must also be noted that the pure mental originality works by the canalization of relevance arising from the primordial nature of God. Thus an originality in the temporal world is conditioned, though not determined, by an initial subjective aim supplied by the ground of all order and of all originality.

Finally, we have to consider the type of structured society which gives rise to the traditional body-mind problem. For example, human mentality is partly the outcome of the human body, partly the single directive agency of the body, partly a system of cogitations which have a certain irrelevance to the physical relationships of the body. The Cartesian philosophy is based upon the seeming fact—the plain fact—of one body and one mind, which are two substances in causal association. For the philosophy of organism the problem is transformed.

The disastrous separation of body and mind, characteristic of philosophical systems which are in any important respect derived from Cartesianism, is avoided in the philosophy of organism by the doctrines of hybrid physical feelings and of the transmuted feelings. In these ways conceptual feelings pass into the category of physical feelings. Also conversely, physical feelings give rise to conceptual feelings, and conceptual feelings give rise to other conceptual feelings. Each actuality is essentially bipolar, physical and mental, and the physical inheritance is essentially accompanied by a conceptual reaction partly conformed to it and partly introductory of a relevant novel contrast, but always introducing emphasis, valuation, and purpose. The integration of the physical and mental side into a unity of experience is a self-formation which is a process of concrescence, and which by the principle of objective immortality characterizes the creativity which transcends it. So though mentality is non-spatial, mentality is always a reaction from, and integration with, physical experience which is spatial.

It is obvious that we must not demand another mentality presiding over these other actualities (a kind of Uncle Sam, over and above all the U.S. citizens). All the life in the body is the life of the individual cells. There are thus millions upon millions of centres of life in each animal body. So what needs to be explained is not dissociation of personality but unifying control, by reason of which we not only have unified behaviour, which can be observed by others, but also consciousness of a unified experience.

A good many actions do not seem to be due to the unifying control, e.g., with proper stimulants a heart can be made to go on beating after it has been taken out of the body. There are centres of reaction and control which cannot be identified with the centre of experience. This is still more so with insects. For example, worms and jellyfish seem to be merely harmonized cells, very little centralized; when cut in two, their parts go on performing their functions independently. Through a series of animals we can trace a progressive rise into a centrality of control. Insects have some central control; even in man, many of the body's actions are done with some independence, but with an organ of central control of very high-grade character in the brain.

The state of things, according to the philosophy of organism, is very different from the Scholastic view of St. Thomas Aquinas, of the mind as informing the body. The living body is a co-ordination of high-grade actual occasions; but in a living body of a low type the occasions are much nearer to a democracy. In a living body of a high type there are grades of occasions so coordinated by their paths of inheritance through the body, that a peculiar richness of inheritance is enjoyed by various occasions in some parts of the body. Finally, the brain is coordinated so that a peculiar richness of inheritance is enjoyed now by this and now by that part; and thus there is produced the presiding personality at that moment in the body. Owing to the delicate organization of the body, there is a returned influence, an inheritance of character derived from the presiding occasion and modifying the subsequent occasions through the rest of the body.

We must remember the extreme generality of the notion of an

enduring object—a genetic character inherited through a historic route of actual occasions. Some kinds of enduring objects form material bodies, others do not. But just as the difference between living and non-living occasions is not sharp, but more or less, so the distinction between an enduring object which is an atomic material body and one which is not, is again more or less. Thus the question as to whether to call an enduring object a transition of matter or of character is very much a verbal question as to where you draw the line between the various properties (cf. the way in which the distinction between matter and radiant energy has now vanished).

Thus in an animal body the presiding occasion, if there be one, is the final node, or intersection, of a complex structure of many enduring objects. Such a structure pervades the human body. It is by reason of the body, with its miracle of order, that the treasures of the past environment are poured into the living occasion. The harmonized relations of the parts of the body constitute this wealth of inheritance into a harmony of contrasts, issuing into intensity of experience. The inhibitions of opposites have been adjusted into the contrasts of opposites. The human mind is thus conscious of its bodily inheritance. There is also an enduring object formed by the inheritance from presiding occasion to presiding occasion. This endurance of the mind is only one more example of the general principle on which the body is constructed.

The final percipient route of occasions is perhaps some thread of happenings wandering in 'empty' space amid the interstices of the brain. It toils not, neither does it spin. It receives from the past; it lives in the present. It is shaken by its intensities of private feeling, adversion or aversion. In its turn, this culmination of bodily life transmits itself as an element of novelty throughout the avenues of the body. Its sole use to the body is its vivid originality: it is the organ of novelty.

Chapter Five

PERCEPTION

The first three chapters have presented Whitehead's analysis of actual occasions. In those early chapters few references were made to traditional philosophical problems; it was necessary, first, to provide the basis for a clear understanding of the basic categories of the Whiteheadian scheme. Chapter Four moved from the level of the microcosmic actual entities to the level of macrocosmic societies such as electrons, cells, trees, and men. In the course of that chapter the manner in which Whitehead's philosophy reconceives, reformulates, and resolves the traditional mind-body problem emerged in the course of the exposition. Here in the present chapter also, the further systematic development of Whitehead's categories focuses attention on his handling of another traditional set of philosophical problems, namely, those connected with perception.

Three concepts are at the heart of Whitehead's theory of perception: perception in the mode of causal efficacy; perception in the mode of presentational immediacy; and perception in the mixed mode of symbolic reference. The first three sections of this chapter deal, respectively, with these three modes of perception.

A few introductory comments will illuminate Whitehead's exposition.

Ordinary, everyday conscious human perception of, say, a gray stone is perception in the mixed mode of symbolic reference. To understand such perception and the possibilities of illusion it involves, a careful analysis of the two more primitive modes that compose the mixture is required. These more primitive modes are causal efficacy and presentational immediacy.

Perception in the mode of causal efficacy is the primitive, ubiquitous feature of all reality that has been referred to in earlier chapters as "prehension." It is the basic mode of inheritance of feeling from past data, and the feelings it transmits are vague, massive, inarticulate, and felt as the efficaciousness of the past. This is what Whitehead refers to as "crude" perception, and it arises in the first phase of concrescence as conformal feeling.

Perception in the mode of presentational immediacy, in contrast, is a product of the later, supplemental phases of experience. It is articulate, sharp, and sophisticated, but lacks the massiveness and power of causal efficacy. Pure presentational immediacy is the objectification, not of the past, but of a contemporary region of space as illustrating specific geometrical, extensive relationships.

Whitehead's quarrel with current accounts of perception can now be explained. These accounts emphasize visual perception, where perception in the mode of presentational immediacy dominates. This emphasis is understandable in that the supplemental phases of experience that produce presentational immediacy dominate human experience and subordinate almost completely the role of causal efficacy, or crude perception. But the theory of perception arising from this emphasis has had fatal consequences for philosophy, notably the consequence that Hume, concentrating on presentational immediacy, correctly inferred that his analysis of presentational immediacy revealed no grounds for validating the concept of causation. Hume's error was not in what he inferred from presentational immediacy, but that he argued from a faulty analysis of perception as a whole. If perception is seen to contain the element of causal efficacy as well as the element of presentational immediacy, then Hume's widely influential analysis of causation fails and much of contemporary philosophy must reexamine its basic presuppositions. Section IV of this chapter presents Whitehead's direct rebuttal of Hume's analysis of causation.

Perception in its primary form is consciousness of the causal efficacy of the external world by reason of which the percipient is a concrescence from a definitely constituted datum. Perception, in this primary sense, is perception of the settled world in the past as constituted by its feeling-tones, and as efficacious by reason of those feeling-tones. Perception, in this sense of the term, will be called 'perception in the mode of causal efficacy.'

What we ordinarily term our visual perceptions are the result of the later stages in the concrescence of the percipient occasion. When we register in consciousness our visual perception of a grey stone, something more than bare sight is meant. A 'stone' has certainly a history, and probably a future. But we all know that the mere sight involved, in the perception of the grey stone, is the sight of a grey shape contemporaneous with the percipient, and with certain spatial relations to the percipient, more or less vaguely defined. Thus the mere sight is confined to the illustration of the geometrical perspective relatedness, of a certain contemporary spatial region, to the percipient, the illustration being effected by the mediation of 'grey.' The sensum 'grey' rescues that region from its vague confusion with other regions. Perception which merely, by means of a sensum, rescues from vagueness a contemporary spatial region, in respect to its spatial shape and its spatial perspective from the percipient, will be called 'perception in the mode of presentational immediacy.'

It is evident that 'perception in the mode of causal efficacy' is not that sort of perception which has received chief attention in the philosophical tradition. The Greeks started from perception in its most elaborate and sophisticated form, namely, visual perception. In visual perception, crude perception is most completely made over by the originative phases in experience, phases which are especially prominent in human experience. Consciousness only illuminates the more primitive types of prehension so far as these prehensions are still elements in the products of integration. But prehensions in the mode of presentational immediacy are among those prehensions which we enjoy with the most vivid consciousness. Thus those elements of our experience which stand out clearly and distinctly in our consciousness are not its basic facts; they are the derivative modifications which arise in the process.

The consequences of the neglect of this law, that the late derivative elements are more clearly illuminated by consciousness than the primitive elements, have been fatal to the proper analysis of an experient occasion. In fact, most of the difficulties of philosophy are produced by it. Experience has been explained in a thoroughly topsy-turvy fashion, the wrong end first. The Greeks were ignorant of modern physics; but modern philosophers discuss perception in terms of categories derived from the Greeks.

The unravelling of the complex interplay between the two modes of perception—causal efficacy and presentational immediacy—is one main problem of the theory of perception. The interplay between the two modes will be termed 'symbolic reference.' Such symbolic reference is so habitual in human experience that great care is required to distinguish the two modes. The ordinary philosophical discussion of perception is almost wholly concerned with this interplay, and ignores the two pure modes which are essential for its proper explanation.

I

Causal Efficacy

In order to find obvious examples of the pure mode of causal efficacy we must have recourse to the viscera and to memory. On this topic I am content to appeal to Hume. He writes: "But my senses convey to me only the impressions of coloured points, disposed in a certain manner. If *the eye is sensible* of anything further, I desire it may be pointed out to me." (Cf. *Treatise*, Bk. I, Part II, Sect. III. Italics not his.) And again: "It is universally allowed by the writers on optics, that *the eye* at all times sees an equal number of physical points, and that a man on the top of a mountain has no larger an image presented to his senses, than when he is cooped up in the narrowest court or chamber." (Cf. Bk. I, Part III, Sect. IX.)

In each of these quotations Hume explicitly asserts that the *eye* sees. The conventional comment on such a passage is that Hume, for the sake of intelligibility, is using common forms of expression; that he is only really speaking of impressions on the mind; and that in the dim future, some learned scholar will gain reputation by emending 'eye' into 'ego.' The reason for citing the passages is to enforce the thesis that the form of speech is literary and intelligible because it expresses the ultimate truth of animal perception. The ultimate momentary 'ego' has as its datum the 'eye as experiencing such-and-such sights.' The point here to be noticed is the immediate literary obviousness of 'the eye as experiencing such-and-such sights.' This is the very reason why Hume uses the expression in spite of his own philosophy. The conclusion, which the philosophy of organism draws, is that in human experience the fundamental fact of perception is the inclusion, in the datum, of the objectification of an antecedent part of the human body with such-and-such experiences. Hume agrees with this conclusion sufficiently well so as to argue from it, when it suits his purpose. He writes: "I would fain ask those philosophers, who found so much of their reasonings on the distinction of substance and accident, and imagine we have clear ideas of each, whether the idea of substance be derived from the impressions of sensation or reflection? If it be conveyed by our senses, I ask, which of them, and after what manner? If it be perceived by the eyes, it must be a colour; if by the ears, a sound; if by the palate, a taste; and so of the other senses." (Cf. Bk. I, Part I, Sect. VI.)

We can prolong Hume's list: the feeling *of* the stone is *in the hand;* the feeling *of* the food is the ache *in the stomach;* the compassionate yearning is *in the bowels,* according to biblical writers; the feeling of well-being is in the viscera *passim;* ill temper is the emotional tone derivative from the disordered liver.

In this list, Hume's and its prolongation, for some cases—as in sight, for example—the supplementary phase in the ultimate subject overbalances in importance the datum inherited from the eye. In other cases, as in touch, the datum of 'the feeling in the hand' maintains its importance, however much the intensity, or even the character, of the feeling may be due to supplementation in the ultimate subject: this instance should be contrasted with that of sight. In the instance of the ache the stomach, as datum, is of chief importance, and the food though obscurely felt is secondary—at least, until the intellectual analysis of the situation due to the doctor, professional or amateur. In the instances of compassion, well-being, and ill temper, the supplementary feelings in the ultimate subject predominate, though there are obscure references to the bodily organs as inherited data.

This survey supports the view that the predominant basis of perception is perception of the various bodily organs, as passing on their experiences by channels of transmission and of enhancement. According to this interpretation, the human body is to be conceived as a complex 'amplifier'—to use the language of the technology of electromagnetism. The enduring personality is the historic route of living occasions which are severally dominant in the body at successive instants. The various actual entities, which compose the body, are so coordinated that the experiences of any part of the body are transmitted to one or more central occasions to be inherited with enhancements accruing upon the way, or finally added by reason of the final integration. The crude aboriginal character of direct perception is inheritance. What is inherited is feeling-tone with evidence of its origin: in other words, vector feeling-tone.

II

Presentational Immediacy

The above introductory remarks concerning causal efficacy will be supplemented by the discussion of symbolic reference in Section III. At the moment a parallel introductory statement concerning presentational immediacy is required.

Presentational immediacy presupposes the notion of an extensive continuum. It will be recalled from Section II of Chapter Four that the widest society discernible, the outermost "Chinese box," is the society constituted by the most general sort of order conceivable, namely, pure extensiveness. This most general of all societies lays down the obligation on everything that is that it conform to this general sort of social order. Another way of putting this is to say that this purely extensive sort of social order limits the perfectly general possibilities of the realm of eternal objects— the only possibilities that are real possibilities in the actual world are those that are compatible with the requirements of this special sort of extensive orderliness, requirements laid down by the past functioning as objectively immortal. These requirements define a continuum, the extensive continuum, within which everything that can be must find its niche. This notion of an extensive continuum is essential to the account of presentational immediacy, hence it is introduced at the beginning.

'Presentational immediacy' presupposes two metaphysical assumptions:

(i) That the actual world, in so far as it is a community of entities which are settled, actual, and already become, conditions and limits the potentiality for creativeness beyond itself. This 'given' world provides determinate data in the form of those objectifications of themselves which the characters of its actual entities can provide. This is a limitation laid upon the general potentiality provided by eternal objects, considered merely in respect to the generality of their natures. Thus, relatively to any actual entity, there is a 'given' world of settled actual entities and a 'real' potentiality, which is the datum for creativeness beyond that standpoint.

Thus we have always to consider two meanings of potentiality: (a) the 'general' potentiality, which is the bundle of possibilities, mutually consistent or alternative, provided by the multiplicity of eternal objects, and (b) the 'real' potentiality, which is conditioned by the data provided by the actual world. General potentiality is absolute, and real potentiality is relative to some actual entity, taken as a standpoint whereby the actual world is defined. The satisfaction of each actual entity is an element in the givenness of the universe: it limits boundless, abstract possibility into the particular real potentiality from which each novel concrescence originates.

(ii) The second metaphysical assumption is that the real potentialities relative to all standpoints are coordinated as diverse determinations of one extensive continuum. This extensive continuum is one relational complex in which all potential objectifications find their niche. It underlies the whole world, past, present, and future. Considered in its full generality, apart from the additional conditions proper only to the cosmic epoch of electrons, protons, molecules, and star-systems, the properties of this continuum are very few and do not include the relationships of metrical geometry. An extensive continuum is a complex of entities united by the various allied relationships of whole to part, and of overlapping so as to possess common parts, and of contact, and of other relationships derived from these primary relationships. This extensive continuum expresses the solidarity of all possible standpoints throughout the whole process of the world. It is not a fact prior to the world; it is the first determination of order—that is, of real potentiality—arising out of the general character of the world.

This extensive continuum is 'real,' because it expresses a fact derived from the actual world and concerning the contemporary actual world. All actual entities are related according to the determinations of this continuum; and all possible actual entities in the future must exemplify these determinations in their relations with the already actual world. The extensive continuum is that general relational element in experience whereby the actual entities experienced, and that unit experience itself, are united in the solidarity of one common world. Extension, apart from its spatialization and temporalization, is that general scheme of relationships providing the capacity that many objects can be welded into the real unity of one experience. Thus, an act of experience has an objective scheme of extensive order by reason of the double fact that its own perspective standpoint has extensive content, and that the other actual entities are objectified with the retention of their extensive relationships. These extensive relations do not make determinate *what* is transmitted; but they do determine conditions to which all transmission must conform. They represent the systematic scheme which is involved in the real potentiality from which every actual occasion *arises*. This scheme is also involved in the attained fact which every actual occasion *is*. The 'extensive scheme' is nothing else than the generic morphology of the internal relations which bind the actual occasions into a nexus, and which bind the prehensions of any one actual occasion into a unity.

*   *   *

We must consider the perceptive mode in which there is clear, distinct consciousness of the 'extensive' relations of the world. These relations include the 'extensiveness' of space and the 'extensiveness' of time. Undoubtedly, this clarity, at least in regard to space, is obtained only in ordinary perception through the senses. This mode of perception is here termed 'presentational immediacy.' In this 'mode' the contemporary world is consciously prehended as a continuum of extensive relations.

So far as physical relations are concerned, contemporary events happen in *causal* independence of each other. This principle lies on the surface of the fundamental Einsteinian formula for the physical continuum. It receives an exemplification in the character of our perception of the world of contemporary actual entities. By reason of the principle of contemporary independence, the contemporary world is objectified for us under the aspect of passive potentiality. Our direct perception of the contemporary world is thus reduced to extension, defining (i) our own geometrical perspectives, and (ii) possibilities of mutual perspectives for other contemporary entities *inter se,* and (iii) possibilities of division. These possibilities of division constitute the external world a continuum, for a continuum is divisible. So far as the contemporary world is divided by actual entities, it is not a continuum, but is atomic. Thus the contemporary world is perceived with its potentiality for extensive division, and not in its actual atomic division.

The contemporary world as perceived by the senses is therefore continuous—divisible but not divided. The contemporary world is in fact divided and atomic, being a multiplicity of definite actual entities. These contemporary actual entities are divided from each other, and are not themselves divisible into other contemporary actual entities. They do, in fact, atomize this continuum; but the aboriginal potentiality, which they include and realize, is what they contribute [to presentational immediacy] as the relevant factor in their objectifications. They thus exhibit the community of contemporary actualities as a common world with mathematical relations.

Presentational immediacy illustrates the contemporary world in respect to its potentiality for extensive subdivision into atomic actualities and in respect to the scheme of perspective relationships which thereby eventuates. But it gives no information as to the actual atomization of this contemporary 'real potentiality.' By its limitations it exemplifies the doctrine, already stated above, that the contemporary world happens independently of the actual occasion with which it is contemporary. This is in fact the definition of contemporaneousness; namely, that actual occasions, *A* and *B*, are mutually contemporary, when *A* does not contribute to the datum for *B*, and *B* does not contribute to the datum for *A*, except that both *A* and *B* are atomic regions in the potential scheme of spatio-temporal extensiveness which is a datum for both *A* and *B*.

The indifference of presentational immediacy to contemporary actualities in the environment cannot be exaggerated. It is only by reason of the fortunate dependence of the experient and of these contemporary actualities on a common past, that presentational immediacy is more than a barren aesthetic display. It does display something, namely, the real extensiveness of the contemporary world. It involves the contemporary actualities but only objectifies them as conditioned by extensive relations. It displays a system pervading the world, a world including and transcending the experient. It is a vivid display of systematic real potentiality, inclusive of the experient and reaching beyond it.

Hume's polemic respecting causation is, in fact, one prolonged, convincing argument that pure presentational immediacy does not disclose any causal influence, either whereby one actual entity is constitutive of the percipient actual entity, or whereby one perceived actual entity is constitutive of another perceived actual entity. The conclusion is that, in so far as concerns their disclosure by presentational immediacy, actual entities in the contemporary universe are causally independent of each other.

Whitehead is, of course, only agreeing with Hume insofar as the latter's analysis is taken as limited to presentational immediacy; in Section IV Whitehead will argue that if taken as a total analysis of perception Hume's views are grossly inadequate.

The paragraphs that follow constitute a slight digression from the main line of the argument. They present a more careful analysis of contemporaneousness than the above and are included because they lead to an intuitive insight into the manner in which Whitehead's system encompasses the implications of modern relativity theory. Whitehead's statements are presented first and are followed by a diagram and commentary that draw out their impli-cations.

The two pure modes of perception disclose a variety of loci de-fined by reference to the percipient occasion M. For example, there are the actual occasions of the settled world which provide the datum for M; these lie in M's causal past. Again, there are the po-tential occasions for which M decides its own potentialities of con-tribution to their data; these lie in M's causal future. There are also those actual occasions which lie neither in M's causal past, nor in M's causal future. Such actual occasions are called M's 'con-temporaries.' These three loci are defined solely by reference to the pure mode of causal efficacy.

We now turn to the pure mode of presentational immediacy. One great difference from the previous way of obtaining loci at once comes into view. In considering the causal mode, the past and the future were defined positively, and the contemporaries of M were defined negatively as lying neither in M's past nor in M's future. In dealing with presentational immediacy the opposite way must be taken. For presentational immediacy gives positive in-formation only about the immediate present as defined by itself. Presentational immediacy illustrates, by means of sensa, potential subdivisions within a cross-section of the world, which is in this way objectified for M. This cross-section is M's immediate present. What is in this way illustrated is the potentiality for subdivision into actual atomic occasions.

Hume's polemic respecting causation constitutes a proof that M's 'immediate present' lies within the locus of M's contemporaries. The presentation to M of this locus, forming its immediate present, contributes to M's datum two facts about the universe: one fact is that there is a 'unison of becoming,' constituting a positive re-lation of all the occasions in this community to any one of them. The members of this community share in a common immediacy; they are in 'unison' as to their becoming: that is to say, any pair of occasions in the locus are contemporaries. The other fact is the subjective illustration of the potential extensive subdivision with complete vagueness respecting the actual atomization. For example, the stone, which in the immediate present is a group of many actual occasions, is illustrated as one grey spatial region. But, to go back to the former fact, the many actual entities of the present stone and the percipient are connected together in the 'unison of immediate becoming.' This community of concrescent occasions, forming M's immediate present, thus establishes a principle of common relatedness, a principle realized as an element in M's datum. This is the principle of mutual relatedness in the 'unison of becoming.' We thus gain the conception of a locus in which any two atomic actualities are in 'concrescent unison,' and which is particularized by the fact that M belongs to it, and so do all actual occasions belonging to extensive regions which lie in M's immediate present. This complete region is the prolongation of M's immediate present beyond M's direct perception, the prolongation being effected by the principle of 'concrescent unison.'

A complete region, satisfying the principle of 'concrescent unison,' will be called a 'duration.' A duration is a cross-section of the universe; it is the immediate present condition of the world at some epoch, according to the old 'classical' theory of time—a theory never doubted until within the last few years. It will have been seen that the philosophy of organism accepts and defines this notion. Some measure of acceptance is imposed upon metaphysics. If the notion be wholly rejected no appeal to universal obviousness of conviction can have any weight; since there can be no stronger instance of this force of obviousness.

The 'classical' theory of time tacitly assumed that a duration included the directly perceived immediate present of each one of its members. The converse proposition certainly follows from the account given above, that the immediate present of each actual occasion lies in a duration. An actual occasion will be said to be 'cogredient with' or 'stationary in' the duration including its directly perceived immediate present. The actual occasion is included in its own immediate present; so that each actual occasion through its percipience in the pure mode of presentational immediacy—if such percipience has important relevance—defines one duration in which it is included. The percipient occasion is 'stationary' in this duration.

But the classical theory also assumed the converse of this statement. It assumed that any actual occasion only lies in one duration;

so that if *N* lies in the duration including *M's* immediate present, then *M* lies in the duration including *N's* immediate present. The philosophy of organism, in agreement with recent physics, rejects this conversion; though its holds that such rejection is based on scientific examination of our cosmic epoch, and not on any more general metaphysical principle. According to the philosophy of organism, in the present cosmic epoch only one duration includes all *M's* immediate present; this one duration will be called *M's* 'presented duration.' But *M* itself lies in many durations; each duration including *M* also includes some portions of *M's* presented duration. In the case of human perception practically all the important portions are thus included.

To sum up this discussion. In respect to any one actual occasion *M* there are three distinct nexūs of occasions to be considered:

(i) The nexus of *M's* contemporaries, defined by the characteristic that *M* and any one of its contemporaries happen in causal independence of each other.

(ii) Durations including *M;* any such duration is defined by the characteristic that any two of its members are contemporaries. (It follows that any member of such a duration is contemporary with *M,* and thence that such durations are all included in the locus (i). The characteristic property of a duration is termed 'unison of becoming.')

(iii) *M's* presented locus, which is the contemporary nexus perceived in the mode of presentational immediacy, with its regions defined by sensa. It is assumed, on the basis of direct intuition, that *M's* presented locus is closely related to some one duration including *M.* It is also assumed, as the outcome of modern physical theory, that there is more than one duration including *M.* The single duration which is so related to *M's* presented locus is termed '*M's* presented duration.'

Figure 4 illustrates these distinctions. The solid lines intersecting at M *delimit the nexus of* M's *contemporaries; any occasion falling in the wedges formed by these solid lines above and below* M *is a contemporary of* M—*i.e., it occurs in causal independence of* M. *In particular,* N *and* P *are both contemporaries of* M.

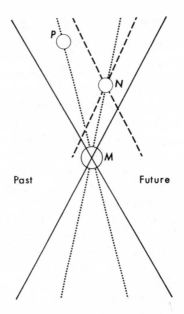

Figure 4. Contemporaneousness.

Within this region of contemporaneousness fall any number of durations including M. The dotted line running through P and M and the dotted line running through N and M represent two such durations. The dash lines intersecting at N delimit the nexus of N's contemporaries. It is crucial to note that P is not a contemporary of N. Therefore there could not be a duration including M and both N and P, because, by (ii) above, a duration is "defined by the characteristic that any two of its members are contemporaries." It follows from (ii) also that while M and P are in unison of becoming and M and N are in unison of becoming, P and N cannot be in unison of becoming.

This denial sounds strange to common sense, but it is precisely this denial that is demanded by relativity theory. In the classical theory, with its assumption that "any actual occasion only lies in one duration," it followed from the definition of a duration that any two occasions sharing a duration with M had to be in the same duration with each other. Admitting many durations through M exactly meets the demands of relativity theory, but the implication as regards the relationship between P and N is an affront to common sense.

An example commonly used to give an intuitive insight into relativity theory to the nonphysicist is as follows. Assume a ray of light to be streaking upward from Cape Kennedy. Relative to Cape Kennedy, the ray of light is traveling at the speed of 186,000 miles per second. Now suppose a rocket to be hurtling upward through the air above Cape Kennedy at a speed of 1,000 miles per second relative to Cape Kennedy. The question now is, what is the speed of the ray of light relative to the rocket? Common sense immediately answers 185,000 miles per second. But relativity theory assures us that this answer is wrong. The upshot of Einstein's theorizing is that the relative velocity of light with respect to any object is always 186,000 miles per second, so that the proper answer is that the speed of light relative to the rocket is also 186,000 miles per second. Now if this is indeed true—and there is the strongest theoretical and experimental evidence that it is—common sense can only conclude that space and time must warp and distort themselves in the vicinity of Cape Kennedy, the rocket, and the ray of light. And this is indeed what relativity asserts, namely that space and time are not absolute, but are liable to distortions.

To return to Whitehead, Figure 4 illustrates this same point. If M is an actual occasion at Cape Kennedy, P an actual occasion contemporary with M in the historic route of the ray of light, and N an actual occasion contemporary with M in the historic route of the rocket, Einstein's transformation equations demand that P and N not be contemporaries—if they were, the relative velocity of the rocket and the ray of light would have to be 185,000 miles per second and not the 186,000 miles per second the theory demands that they be.

But there is one final point of considerable importance to be noted in these paragraphs, a point that brings the discussion back to the topic of presentational immediacy. Whitehead notes that the conviction of the "old 'classical' theory of time" that there is an "immediate present condition of the world at some epoch" is based on such "obviousness of conviction" that its acceptance in some measure "is imposed upon metaphysics." Whitehead's "measure of acceptance" of the classical theory is expressed in terms of pres-

entational immediacy. An actual entity objectifies one of the many durations in which it is included when it perceives in the mode of presentational immediacy. This duration is consequently the one associated with that actual entity's "directly perceived immediate present," and the subsequent overpowering influence of that duration accounts for the "obviousness of conviction" associated with the classical theory. In this way Whitehead accounts for the intuitive plausibility of the classical theory while at the same time he incorporates into his philosophy the discoveries of modern relativity theory.[1]

III

Symbolic Reference

Perception in the mode of causal efficacy is awareness of the massive presence of the past. Perception in the mode of presentational immediacy is awareness of the extensive relationships structuring a continuum experienced as constituting the immediate present. Neither of these modes is what is experienced as ordinary perception; ordinary perception is a mixed mode of perception compounded out of the two more primitive modes. This present section explains how the two primitive modes interact to produce the mixed mode of symbolic reference.

Figure 5 indicates the essential aspect of this interaction. The final paragraph in Whitehead's exposition (paragraph iii of the last material quoted) noted that "the contemporary nexus perceived in the mode of presentational immediacy" is perceived "with its regions defined by sensa." The crux of symbolic reference is that the sensa that define regions in the contemporary nexus are not given in the mode of presentational immediacy, but are, rather, donated from the mode of causal efficacy and "projected" onto the contemporary nexus revealed in presentational immediacy. Figure 5 schematizes this interaction. N is in the causal past of M and M

---

[1] Whitehead refines upon these distinctions even further in a very technical discussion of the definition of a straight line and the consequences of such a definition. These further distinctions are not required for present purposes, but the interested reader is referred to Part IV of *PR*, and in particular to Chapter IV of Part IV.

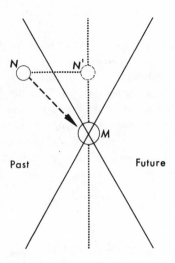

Past

Future

Figure 5. Symbolic Reference.

inherits from N *in the mode of causal efficacy.* M *also is aware, via presentational immediacy, of the extensive characteristics of the continuum in its presented duration (dotted line), a continuum in fact atomized in a certain region by* N′. *If* M *is not the victim of illusion, it will project the sensa inherited from* N *in the mode of causal efficacy onto the region of the extensive continuum in fact atomized by* N′. *If* N *is a book on* M's *desk, the chance for error is small; if* N *is a star ten million light years away that exploded five million years ago,* M *still projects, but erroneously, sensa derived from* N *onto its presented duration. The greatest source of error, however, arises from the fact that the final portion of the route from* N *to* M *is through* M's *body up to the historic route of regnant, living occasions severally dominant in the body at successive instants. If the body be tired, drugged, or otherwise handicapped, the effect on symbolic reference can be catastrophic.*

The pure mode of presentational immediacy gives no information as to the past or the future. It merely presents an illustrated portion of the presented duration. It thereby defines a cross-section of the universe: but does not in itself define on which side lies the past, and on which side the future. In order to solve such questions we now come to the interplay between the two pure modes. This mixed mode of perception is here named 'symbolic reference.' The failure to lay due emphasis on symbolic reference is one of the reasons for metaphysical difficulties; it has reduced the notion of 'meaning' to a mystery. The first principle, explanatory of symbolic reference, is that for such reference a 'common ground' is required. By this necessity for a 'common ground' it is meant that there must be components in experience which are directly recognized as identical in each of the pure perceptive modes.

One main element of common ground, shared between the two pure modes, is the presented locus. This locus enters subordinately into the perceptive mode of causal efficacy, vaguely exemplifying its participation in the general scheme of extensive interconnection, involved in the real potentiality. It is not disclosed by that perceptive mode in any other way; at least it is not directly disclosed. The further disclosure must be indirect, since contemporary events are exactly those which are neither causing, nor caused by, the percipient actual occasion. Now, although the various causal pasts (i.e. 'actual worlds') of the contemporary actual occasions are not wholly identical with the causal past of the percipient actual occasion, yet, so far as important relevance is concerned, these causal pasts are practically identical. Thus there is, in the mode of causal efficacy, a direct perception of those antecedent actual occasions which are causally efficacious both for the percipient and for the relevant events in the presented locus.[2] The percipient therefore, under the limitation of its own perspective, prehends the causal influences to which the presented locus in its important regions is subjected. This amounts to an indirect perception of this locus, a perception in which the direct components belong to the pure mode of causal efficacy.

If we now turn to the perceptive mode of presentational immediacy, the regions, perceived by direct and indirect knowledge respectively, are inverted in comparison with the other mode. The presented locus is directly illustrated by the sensa; while the causal past, the causal future, and the other contemporary events, are only indirectly perceived by means of their extensive relations to

[2] In terms of Figure 5, N is efficacious both for M and for N'.

the presented locus. The presented locus, with the animal body of the percipient as the region from which perspectives are focussed, is the regional origin by reference to which in this perceptive mode the complete scheme of extensive regions is rendered determinate.

The presented locus is a common ground for the symbolic reference, because it is directly and distinctly perceived in presentational immediacy, and is indistinctly and indirectly perceived in causal efficacy. In the latter mode, the indistinctness is such that the detailed geometrical relationships are, for the most part, incurably vague. Particular regions are, in this perceptive mode, in general not distinguishable. In this respect, causal efficacy stands in contrast to presentational immediacy with its direct illustration of certain distinct regions.

The second 'ground' for symbolic reference is the connection between the two modes effected by the identity of an eternal object ingredient in both of them. The bare mathematical potentialities of the extensive continuum require an additional content in order to assume the rôle of real objects for the subject. This content is supplied by the eternal objects termed sense-data. These objects are 'given' for the experience of the subject. Their givenness does not arise from the 'decision' of the contemporary entities which are thus objectified. It arises from the functioning of the antecedent physical body of the subject; and this functioning can in its turn be analysed as representing the influence of the more remote past, a past common alike to the subject and to its contemporary actual entities. Thus these sense-data are eternal objects playing a complex relational rôle; they connect the actual entities of the past with the actual entities of the contemporary world, and thereby effect objectifications of the contemporary things and of the past things.

In tracing the origin of presentational immediacy, we find mental operations transmuting the functions of sensa so as to transfer them from being participants in causal prehensions into participants in presentational prehensions. In respect to the sensa concerned, there is a gradual transformation of their functions as they pass from occasion to occasion along a route of inheritance up to some final high-grade experient. In their most primitive form of functioning, a sensum is felt physically with emotional enjoyment of its sheer individual essence. For example, red is felt with emotional enjoyment of its sheer redness. In this primitive prehension we have aboriginal physical feeling in which the subject feels itself as enjoying redness. This is the most primitive form of the feeling of causal efficacy. In physics it is the transmission of a form of energy. In the bodily transmission from occasion to occasion of a high-grade animal body, there is a gradual modification of these functions of sensa. In their most primitive functioning for the initial occasions within the animal body, they are qualifications of emotion—types of energy, in the language of physics: in their final functioning for the high-grade experient occasion at the end of the route, they are qualities 'inherent' in a presented, contemporary nexus.

Thus a sense-datum has ingression into experience by reason of its forming the *what* of a very complex multiple integration of prehensions within that occasion. For example, the ingression of a visual sense-datum involves the causal objectification of various antecedent bodily organs and the presentational objectification of the shape seen, this shape being a nexus of contemporary actual entities.

Presentational immediacy is the enhancement of the importance of relationships which were already in the datum, vaguely and with slight relevance. This fact, that 'presentational immediacy' deals with the same datum as does 'causal efficacy,' gives the [second] reason why there is a common 'ground' for 'symbolic reference.' The two modes express the same datum under different proportions of relevance.

Thus symbolic reference, though in complex human experience it works both ways, is chiefly to be thought of as the elucidation of percepta in the mode of causal efficacy by the fluctuating intervention of percepta in the mode of presentational immediacy.

The former mode produces percepta which are vague, not to be controlled, heavy with emotion: it produces the sense of derivation from an immediate past, and of passage to an immediate future; a sense of emotional feeling, belonging to oneself in the past, passing into oneself in the present, and passing from oneself in the present towards oneself in the future; a sense of influx of influence from other vaguer presences in the past, localized and yet evading local definition, such influence modifying, enhancing, inhibiting, diverting, the stream of feeling which we are receiving, unifying, enjoying, and transmitting. This is our general sense of existence, as one item among others, in an efficacious actual world.

The percepta in the mode of presentational immediacy have the converse characteristics. In comparison, they are distinct, definite, controllable, apt for immediate enjoyment, and with the minimum of reference to past, or to future. Presentational immediacy is an outgrowth from the complex datum implanted by causal efficacy. But, by the originative power of the supplemental phase, what was vague, ill defined, and hardly relevant in causal efficacy, becomes distinct, well defined, and importantly relevant in presentational immediacy. The supplemental phase [via presentational immediacy] lifts the presented duration into vivid distinctness, so that the vague efficacy of the indistinct external world in the immediate past is precipitated upon the representative regions in the contemporary present. In the usual language, the sensations are projected.

We are subject to our percepta in the mode of efficacy, we adjust our percepta in the mode of immediacy. But, in fact, our process of self-construction for the achievement of unified experience produces a new product, in which percepta in one mode, and percepta in the other mode, are synthesized into one subjective feeling. For example, we are perceiving before our eyes a grey stone. The two modes are unified by a blind symbolic reference by which supplemental feelings derived from the intensive, but vague, mode of efficacy are precipitated upon the distinct regions illustrated in the mode of immediacy. The integration of the two modes in supplemental feeling makes what would have been vague to be distinct, and what would have been shallow to be intense. This is the perception of the grey stone, in the mixed mode of symbolic reference.

Generally—though not always—the adjectival words express information derived from the mode of immediacy, while the substantives convey our dim percepts in the mode of efficacy. For example, 'grey' refers to the grey shape immediately before our eyes: this percept is definite, limited, controllable, pleasant or unpleasant, and with no reference to past or to future. It is this sort of percept which has led to Descartes' definition of substances as 'requiring nothing but themselves in order to exist,' and to his notion of 'extension' as the principal attribute of a genus of substances. It has also led to Hume's notion of 'impressions of sensation' arising from unknown sources, and in complete independence so far as any discernible nexus is concerned. But the other element in the compound percept has a widely different character. The word 'stone' is selected, no doubt, because its dictionary meaning will afford some help in understanding the particular percepta meant. But the word is meant to refer to particular feelings of efficacy in the immediate past, combined with anticipations for the immediate future; this feeling is vaguely localized, and conjecturally identified with the very definite localization of the 'grey' perceptum.

Thus, so far as concerns conscious judgment, the symbolic reference is the acceptance of the evidence of percepta, in the mode of immediacy, as evidence for the localization and discrimination of vague percepta in the mode of efficacy. So far as bodily feelings are concerned, there is some direct check on this procedure; but, beyond the body, the appeal is to the pragmatic consequences,, involving some future state of bodily feelings which can be checked up.

Symbolic reference belongs to one of the later originative phases of experience. These later phases are distinguished by their new element of originative freedom. Accordingly, while the two pure perceptive modes are incapable of error, symbolic reference introduces this possibility. When human experience is in question, 'perception' almost always means 'perception in the mixed mode of symbolic reference.' Thus, in general, human perception is subject to error, because, in respect to those components most clearly in consciousness, it is interpretative. In fact, error is the mark of the higher organisms, and is the schoolmaster by whose agency there is upward evolution. For example, the evolutionary use of intelligence is that it enables the individual to profit by error without being slaughtered by it.

Perception can be erroneous, in the sense that the feeling associates regions in the presented locus with inheritances from the past, which in fact have not been thus transmitted into the present regions. In the mixed mode, the perceptive determination is purely due to the bodily organs, and thus there is a gap in the perceptive logic—so to speak. This gap is not due to any conceptual freedom

on the part of the ultimate subject. It is not a mistake due to consciousness. It is due to the fact that the body, as an instrument for
synthesizing and enhancing feelings, is faulty, in the sense that it
produces feelings which have but slight reference to the real state
of the presented duration. The projection of sensa in presentational immediacy depends entirely upon the state of the brain and
upon systematic geometrical relations characterizing the brain.
How the brain is excited, whether by visual stimuli through the
eye, or by auditory stimuli through the ear, or by the excessive
consumption of alcohol, or by hysterical emotion, is completely indifferent; granted the proper excitement of the brain, the experient
will perceive some definite contemporary region illustrated by the
projected sensa. When we see a coloured shape, it may be a real
man, or a ghost, or an image behind a mirror, or an hallucination;
but whatever it be, *there* it *is*—exhibiting to us a certain region of
external space. The word 'delusive' is all very well as a technical
term; but it must not be misconstrued to mean that what we *have*
directly perceived, we have *not* directly perceived. Our direct perception, via our senses, of an immediate extensive shape, in a
certain geometrical perspective to ourselves, and in certain general
geometrical relations to the contemporary world, remains an ultimate fact. Our inferences are at fault.

The question as to which regions have their relatedness to other
constituents of the datum—such as 'grey,' for instance—depends
upon the coordination of the bodily organs through which the
routes of inheritance pass. In a fortunately constructed animal
body, this selection is determined chiefly by the inheritance received by the superficial organs—the skin, the eyes, etc.—from
the external environment, and preserves the relevance of the vector
character of that external inheritance. When this is the case, the
perceptive mode of immediacy has definite relevance to the future
efficacy of the external environment, and then indirectly illustrates
the inheritance which the presented locus receives from the immediate past. The animal body is so constructed that, with rough
accuracy and in normal conditions, important emphasis is thus
laid upon those regions in the contemporary world which are
particularly relevant for the future existence of the enduring object of which the immediate percipient is one occasion.

IV
Perception, Causation, and Whitehead's Rebuttal of Hume

In this section the distinctions made in the earlier parts of the chapter are applied to one of the basic problems of philosophy, the problem of causation. This problem was elevated to its present position of importance by the analysis of David Hume. Whitehead directly attacks Hume, arguing as follows: (1) Hume's theory of causation depends on his theory of perception; (2) that theory of perception is inadequate; consequently (3) his analysis of causation is inadequate.

The discussion of the problem, constituted by the connection between causation and perception, has been conducted by the various schools of thought derived from Hume and Kant under the misapprehension generated by an inversion of the true constitution of experience. The inversion was explicit in the writings of Hume and of Kant: for both of them presentational immediacy was the primary fact of perception, and any apprehension of causation was, somehow or other, to be elicited from this primary fact. This view of the relation between causation and perception, as items in experience, was not original to these great philosophers. It is to be found presupposed in Locke and Descartes; and they derived it from mediaeval predecessors. But the modern critical movement in philosophy arose when Hume and Kant emphasized the fundamental, inescapable, importance which this doctrine possesses for any philosophy admitting its truth. The philosophy of organism does not admit its truth, and thus rejects the touchstone which is the neolithic weapon of 'critical' philosophy. It must be remembered that clearness in consciousness is no evidence for primitiveness in the genetic process: the opposite doctrine is more nearly true.

Owing to its long dominance, it has been usual to assume as an obvious fact the primacy of presentational immediacy. We open our eyes and our other sense-organs; we then survey the contemporary world decorated with sights, and sounds, and tastes; and then, by the sole aid of this information about the contemporary world, thus decorated, we draw what conclusions we can as to the actual world. No philosopher really holds that this is the sole source of information: Hume and his followers appeal vaguely to 'memory' and to 'practice,' in order to supplement their direct information; and Kant wrote other *Critiques* in order to supplement his *Critique of Pure Reason*. But the general procedure of modern philosophical 'criticism' is to tie down opponents strictly to the front door of presentational immediacy as the sole source of information, while one's own philosophy makes its escape by a back door veiled under the ordinary usages of language.

If this 'Humian' doctrine be true, certain conclusions as to 'behaviour' ought to follow—conclusions which, in the most striking way, are not verified. It is almost indecent to draw the attention of philosophers to the minor transactions of daily life, away from the classic sources of philosophic knowledge; but, after all, it is the empiricists who began this appeal to Caesar.

According to Hume, our behaviour presupposing causation is due to the repetition of associated presentational experiences. Thus the vivid presentment of the antecedent percepts should vividly generate the behaviour, in action or thought, towards the associated consequent. The clear, distinct, overwhelming perception of the one is the overwhelming reason for the subjective transition to the other. For behaviour, interpretable as implying causation, is on this theory the subjective response to presentational immediacy. According to Hume this subjective response is the beginning and the end of all that there is to be said about causation. In Hume's theory the response is response to presentational immediacy, and to nothing else. Also the situation elicited in response is nothing but an immediate presentation, or the memory of one.

Let us apply this explanation to reflex action: In the dark, the electric light is suddenly turned on and the man's eyes blink.

There is a simple physiological explanation of this trifling incident. But this physiological explanation is couched wholly in terms of causal efficacy: it is the conjectural record of the travel of a spasm of excitement along nerves to some nodal centre, and of the return spasm of contraction back to the eyelids. The correct technical phraseology would not alter the fact that the explanation does not involve any appeal to presentational immediacy either for actual occasions resident in the nerves, or for the man. At the most there is a tacit supposition as to what a physiologist, who in fact was not there, might have seen if he had been there, and if he could have vivisected the man without affecting these occurrences, and if he could have observed with a microscope which also in fact was absent. Thus the physiological explanation remains, from the point of view of Hume's philosophy, a tissue of irrelevancies. It presupposes a side of the universe about which, on Hume's theory, we must remain in blank ignorance.

Let us now dismiss physiology and turn to the private experience of the blinking man. The sequence of percepts, in the mode of presentational immediacy, is flash of light, feeling of eye-closure, instant of darkness. The three are practically simultaneous; though the flash maintains its priority over the other two, and these two latter percepts are indistinguishable as to priority. According to the philosophy of organism, the man also experiences another percept in the mode of causal efficacy. He feels that the experiences of the *eye* in the matter of the flash are causal of the blink. The man himself will have no doubt of it. In fact, it is the feeling of causality which enables the man to distinguish the priority of the flash; and the inversion of the argument, whereby the temporal sequence 'flash to blink' is made the premise for the 'causality' belief, has its origin in pure theory. The man will explain his experience by saying, 'The flash made me blink'; and if his statement be doubted, he will reply, 'I know it, because I felt it.'

The philosophy of organism accepts the man's statement, that the flash *made* him blink. But Hume intervenes with another explanation. He first points out that in the mode of presentational immediacy there is no percept of the flash *making* the man blink. In this mode there are merely the two percepts—the flash and the blink—combining the two latter of the three percepts under the one term 'blink.' Hume refuses to admit the man's protestation, that the compulsion to blink is just what he did feel. The refusal is based on the dogma, that all percepts are in the mode of presentational immediacy—a dogma not to be upset by a mere appeal to direct experience. Besides, Hume has another interpretation of the man's experience: what the man really felt was his *habit* of blinking after flashes. The word 'association' explains it all, according to Hume. But how can a 'habit' be felt, when a 'cause' cannot be felt? Is there any presentational immediacy in the feeling of a 'habit'? Hume by a sleight of hand confuses a 'habit of feeling blinks after flashes' with a *'feeling of the habit* of feeling blinks after flashes.' We have here a perfect example of the practice of applying the test of presentational immediacy to procure the critical rejection of some doctrines, and of allowing other doctrines to slip out by a back door, so as to evade the test. The notion of causation arose because mankind lives amid experiences in the mode of causal efficacy.

We will keep to the appeal to ordinary experience, and consider another situation, which Hume's philosophy is ill equipped to explain. The 'causal feeling' according to that doctrine arises from the long association of well-marked presentations of sensa, one precedent to the other. It would seem therefore that inhibitions of sensa, given in presentational immediacy, should be accompanied by a corresponding absence of 'causal feeling'; for the explanation of how there is 'causal feeling' presupposes the well-marked familiar sensa, in presentational immediacy. Unfortunately the contrary is the case. An inhibition of familiar sensa is very apt to leave us a prey to vague terrors respecting a circumambient world of causal operations. In the dark there are vague presences, doubtfully feared; in the silence, the irresistible causal efficacy of nature presses itself upon us; in the vagueness of the low hum of insects in an August woodland, the inflow into ourselves of feelings from enveloping nature overwhelms us; in the dim consciousness of half-sleep, the presentations of sense fade away, and we are left with the vague feeling of influences from vague things around us. It is quite untrue that the feelings of various types of influences are dependent upon the familiarity of well-marked sensa in immediate presentment. Every way of omitting the sensa still leaves us a prey to vague feelings of influence. Such feelings, divorced from immediate sensa, are pleasant, or unpleasant, according to mood; but they are always vague as to spatial and temporal definition, though their explicit dominance in experience may be heightened in the absence of sensa.

Further, our experiences of our various bodily parts are primarily perceptions of them as *reasons* for "projected' sensa: the *hand* is the *reason* for the projected touch-sensum, the *eye* is the *reason* for the projected sight-sensum. Our bodily experience is primarily an experience of the dependence of presentational immediacy upon causal efficacy. Hume's doctrine inverts this relationship by making causal efficacy, as an experience, dependent upon presentational immediacy. This doctrine, whatever be its merits, is not based upon any appeal to experience. Hume's doctrine has no recommendation except the pleasure which it gives to its adherents.

Bodily experiences, in the mode of causal efficacy, are distinguished by their comparative accuracy of spatial definition. The causal influences from the body have lost the extreme vagueness of those which inflow from the external world. But, even for the body, causal efficacy is dogged with vagueness compared to presentational immediacy. These conclusions are confirmed if we descend the scale of organic being. It does not seem to be the sense of causal awareness that the lower living things lack, so much as variety of sense-presentation, and then vivid distinctness of presentational immediacy. But animals, and even vegetables, in low forms of organism exhibit modes of behaviour directed towards self-preservation. There is every indication of a vague feeling of causal relationship with the external world, of some intensity, vaguely defined as to quality, and with some vague definition as to locality. A jellyfish advances and withdraws, and in so doing exhibits some perception of causal relationship with the world beyond itself; a plant grows downwards to the damp earth, and upwards towards the light. There is thus some direct reason for attributing dim, slow feelings of causal nexus, although we have no reason for any ascription of the definite percepts in the mode of presentational immediacy.

As we pass to the inorganic world, causation never for a moment seems to lose its grip. What is lost is originativeness, and any evidence of immediate absorption in the present. Thus we must assign the mode of causal efficacy to the fundamental constitution of an occasion so that in germ this mode belongs even to organisms of the lowest grade; while the mode of presentational immediacy requires the more sophisticated activity of the later stages of process, so as to belong only to organisms of a relatively high grade.

Chapter Six

WHITEHEAD
AND OTHER
PHILOSOPHERS

It is not the purpose of this chapter to study the thought of certain philosophers other than Whitehead. Rather, the intent is to use these other philosophers as a foil with the aim of throwing further light on Whitehead's system. In the passages that follow, Whitehead works his way into the heart of several major positions with the aim of showing both what he accepts and what he rejects of certain classical philosophers. These passages not only serve to illuminate Whitehead's philosophy of organism, they also shed light on Whitehead's relationship to the great tradition of philosophical speculation. To understand this relationship is to understand why Whitehead frequently introduces new, strange-sounding, and initially bewildering terminology in order to avoid traps and pitfalls that other philosophers have stumbled into, and also to understand that, in spite of the strange terminology, Whitehead is frequently closer to traditional positions than his mode of speaking initially suggests.

The following sentences from Whitehead set the tone of this chapter:

A detailed discussion of Descartes, Locke, and Hume may make plain how deeply the philosophy of organism is founded on seventeenth-century thought and how at certain critical points it diverges from that thought. The scheme of interpretation here adopted can claim for each of its main positions the express authority of one, or the other, of some supreme master of thought— Plato, Aristotle, Descartes, Locke, Hume, Kant. But ultimately nothing rests on authority; the final court of appeal is intrinsic reasonableness.

I
Descartes (and Hume)

It is impossible to scrutinize too carefully the character to be assigned to the datum in the act of experience. The whole philosophical system depends on it. Hume's doctrine of 'impressions of sensation' (*Treatise,* Book I, Part I, Sect. II) is twofold. I will call one part of his doctrine 'The Subjectivist Principle' and the other part 'The Sensationalist Principle.' It is usual to combine the two under the heading of the 'sensationalist doctrine'; but two principles are really involved, and many philosophers—Locke, for instance— are not equally consistent in their adhesion to both of them. The philosophy of organism denies both of these doctrines, though it accepts a reformed subjectivist principle.

The subjectivist principle is, that the datum in the act of experience can be adequately analysed purely in terms of universals.

The sensationalist principle is, that the primary activity in the act of experience is the bare subjective entertainment of the datum, devoid of any subjective form of reception. This is the doctrine of *mere* sensation.

Throughout this chapter Whitehead recurs to these two principles again and again, so they must be kept firmly in mind. The subjectivist principle, by asserting that the datum in the act of experience can be exhausted by an analysis into universals, denies that the datum is experienced as determined to particular existents. This denial is anathema to Whitehead; it leads directly to skepticism, as Hume so clearly showed. Whitehead will argue that although Descartes and Locke officially accept the subjectivist principle, there are moments when each repudiates the principle

—*Descartes with his doctrine of* realitas objectiva *and Locke by expanding his concept of an "idea" in Books III and IV of his Essay. These moments are their more profound moments, Whitehead argues, and they foreshadow the doctrine of "objectification" in the philosophy of organism.*

There is an important corollary of the subjectivist principle. If the total datum in the act of experience can be analyzed purely in terms of universals, or sensa, this means that all experience is in the mode of presentational immediacy and all perception is sense perception. This is to ignore perception in the mode of causal efficacy, and such a procedure leads to Hume's erroneous analysis of causation, as was described in Section IV of the previous chapter. To put it another way, the subjectivist principle commits those who accept it to the view that conscious experience is the foundation of experience, and not the culmination of experience as Whitehead would want to insist. It is in terms of this implication of the subjectivist principle that Whitehead argues in Section III of the present chapter that Kant's "Transcendental Aesthetic" is "a distorted fragment of what should have been his main topic" —i.e., Kant should have spent more time talking about the ordering of experience at the preconscious, preintellectual level.

The sensationalist principle is simpler and more straightforward. It is "the doctrine of mere *sensation*"—i.e., the doctrine that the perceiver is passive in the act of experience. Whitehead's doctrine of subjective form repudiates the sensationalist principle as does Kant's philosophy, which is based on the principle that the perceiver contributes the formal element in experience. Whitehead notes that Kant's importance to philosophy is that he realized that "the sensationalist principle acquires dominating importance, if the subjectivist principle be accepted." This realization is the basis for Kant's Copernican Revolution in philosophy.

These two principles, sometimes distinguished and sometimes lumped together as "the sensationalist doctrine" or as "subjective sensationalism," are never far from Whitehead's mind in what follows.

The subjectivist principle follows from three premises: (i) The acceptance of the 'substance-quality' concept as expressing the ultimate ontological principle.[1] (ii) The acceptance of Aristotle's definition of a primary substance, as always a subject and never a predicate. (iii) The assumption that the experient subject is a primary substance. The first premise states that the final metaphysical fact is always to be expressed as a quality inhering in a substance. The second premise divides qualities and primary substances into two mutually exclusive classes. The two premises together are the foundation of the traditional distinction between universals and particulars.

The philosophy of organism denies the premises on which this distinction is founded. It admits two ultimate classes of entities, mutually exclusive. One class consists of 'actual entities,' which in the philosophical tradition are mis-described as 'particulars'; and the other class consists of forms of definiteness, here named 'eternal objects,' which in comparison with actual entities are mis-described as 'universals.' The ontological principle, and the wider doctrine of universal relativity, on which the present metaphysical discussion is founded, blur the sharp distinction between what is universal and what is particular. The notion of a universal is of that which can enter into the description of any other particular. According to the doctrine of relativity which is the basis of the metaphysical system of the present lectures, both these notions involve a misconception. The term 'universal' is unfortunate in its application to eternal objects; for it seems to deny, and in fact it was meant to deny, that the actual entities also fall within the scope of the principle of relativity. An actual entity cannot be described, even inadequately, by universals; because other actual entities do enter into the description of any one actual entity. Thus every so-called 'universal' is particular in the sense of being just what it is, diverse from everything else; and every so-called 'particular' is universal in the sense of entering into the constitutions of other actual entities. The contrary opinion led to the collapse of Descartes' many substances into Spinoza's one substance; to Leibniz's windowless monads with their pre-established harmony; to the sceptical reduction of Hume's philosophy—a reduction first

[1] Here Whitehead merely acknowledges that the subjectivist principle depends upon the substance-quality concept. Later in this section he attacks the substance-quality concept directly.

effected by Hume himself, and reissued with the most beautiful exposition by Santayana in his *Scepticism and Animal Faith.*

The point is that the current view of universals and particulars inevitably leads to the epistemological position stated by Descartes: "From this I should conclude that I knew the wax by means of vision and not simply by the intuition of the mind; unless by chance I remember that, when looking from a window and saying I see men who pass in the street, I really do not see them, but infer that what I see is men, just as I say that I see wax. And yet what do I see from the window but hats and coats which may cover automatic machines? Yet I judge these to be men. And similarly solely by the faculty of judgment (*judicandi*) which rests in my mind, I comprehend that which I saw with my eyes." (Cf. Meditation II.)

In this passage it is assumed[2] that Descartes—the Ego in question—is a particular, characterized only by universals. Thus his impressions—to use Hume's word—are characterizations by universals. Thus there is no perception of a particular actual entity. He arrives at the belief in the actual entity by 'the faculty of judgment.' But on this theory he has absolutely no analogy upon which to found any such inference with the faintest shred of probability. Hume, accepting Descartes' account of perception (in this passage), which also belongs to Locke in some sections of his *Essay,* easily draws the sceptical conclusion. Santayana irrefutably exposes the full extent to which this scepticism must be carried. Descartes held, with some flashes of inconsistency arising from the use of '*realitas objectiva*,' the subjectivist principle as to the datum. But he also held that this mitigation of the subjectivist principle [the mitigation arising from the use of '*realitas objectiva*'] enabled the 'process' within experience to include a sound argument for the existence of God; and thence a sound argument for the general veridical character of those presumptions as to the external world which somehow arise in the process.

According to the philosophy of organism, it is only by the intro-

2 Perhaps inconsistently with what Descartes says elsewhere: in other passages the mental activity involved seems to be *analysis* which discovers '*realitas objectiva*' as a component element of the idea in question. There is thus '*inspectio*' rather than '*judicium*.' [Footnote is Whitehead's.]

duction of covert inconsistencies into the subjectivist principle, as here stated, that there can be any escape from what Santayana calls, 'solipsism of the present moment.' Thus Descartes' mode of escape is either illusory, or its premises are incompletely stated. This covert introduction is always arising because common sense is inflexibly objectivist. The philosophy of organism recurs to Descartes' alternative theory of 'realitas objectiva,' and endeavours to interpret it in terms of a consistent ontology. Descartes endeavoured to combine the two theories; but his unquestioned acceptance of the subject-predicate dogma forced him into a representative theory of perception, involving a 'judicium' validated by our assurance of the power and the goodness of God.

The logical structure of Whitehead's analysis of Descartes' philosophy can be stated simply. Whitehead maintains that Descartes holds an inconsistent position. For the most part Descartes embraces a representative theory of perception, insisting that experience reveals no concrete particulars but only universals, which one then infers to reveal aspects of particulars. But if this is all there is to it, Whitehead argues, the inferences can never be justified and skepticism is inevitable. Of course Descartes wishes to avoid skepticism. He does it, Whitehead maintains, only by inconsistently introducing at one point in his argument an exception to his representative theory, an exception built into what Whitehead refers to as "Descartes' doctrine of 'realitas objectiva'." This doctrine must be understood in order to grasp Whitehead's criticism.

In one of his arguments for the existence of God in the third meditation, Descartes establishes that a cause must have as much reality as its effect. He then notes his idea of God. This is only an idea, but it nevertheless has a very high degree of that kind of reality possessed by objects in the mind, namely "realitas objectiva." (Note that the word objective for Descartes does not have our modern meaning, but has the special medieval meaning of "conceptual" or of being "in the mind.") Descartes then asserts that the cause of this idea of God must have as much "formal" reality (and for Descartes this means reality independent of mind) as the idea of God has "realitas objectiva" (i.e., reality in the mind). *The conclusion is that only God could have formal reality commensurate with the "realitas objectiva" of the idea of God and hence, because the latter does exist, the former must exist.*

It is in this argument, Whitehead maintains, that Descartes has inconsistently repudiated the representative theory of perception. The claim, made by Descartes, that his ideas, considered as images, differ among themselves in that some of them "contain in themselves, so to speak, more objective reality" is, to Whitehead, an example of admitting "covert inconsistencies into the subjectivist principle."

Let it be firmly understood that Whitehead is on the side of the inconsistency here. But instead of slipping it in covertly in his own philosophy, he makes what for Descartes is merely a brilliant inconsistency the cornerstone of his whole metaphysics by holding that an actual entity prehends other actual entities. This is to claim, against the representative theory, that experience reveals not simply universals, but, in Locke's phrase, "universals determined to particulars." Another way of making this claim in the philosophy of organism is to insist that perception in the mode of causal efficacy is a part of experience as well as perception in the mode of presentational immediacy. Descartes, when he is being consistently a representationalist, speaks as though experience were exhausted by pure sense perception (which occurs in the mode of presentational immediacy); in the "realitas objectiva" doctrine he inconsistently introduces an exception to his main position. Whitehead finds the inconsistent Descartes more profound than the consistent Descartes and places the notion of causal efficacy at the center of his philosophy. It is obvious that the argument advanced here is essentially the same as that utilized in Section IV of Chapter Five to refute Hume's analysis of causation, though the justification is of wider scope and placed more explicitly in an historical context.

This commentary on Whitehead's analysis of Descartes applies equally well to his analysis of Locke. In Section II of the present chapter it is argued that the first two books of Locke's Essay *contain essentially the same representative theory of perception that is held by Descartes, but that in Books III and IV Locke introduces in a more profound moment a strain of thought that is inconsistent with his main position and parallel to Descartes' notion of "realitas objectiva."*

Whitehead views Hume as the philosopher who pursues the consistent Descartes and the consistent Locke to the skepticism inherent in their representative theory; he views himself as avoiding skepticism by accepting as central to his philosophy the more profound insights of the inconsistent Descartes and the inconsistent Locke, who more truly express the nature of experience in those unguarded moments when they examine it naively without being shackled by the a priori demands of their representative theory.

Whitehead turns now to another aspect of Descartes' philosophy. Descartes is responsible for the so-called "subjectivist bias" of modern philosophy, and in the paragraphs that follow Whitehead traces briefly the origin of this subjectivist bias and specifies the sense in which it is accepted by his philosophy of organism.

Now philosophy has always proceeded on the sound principle that its generalizations must be based upon the primary elements in actual experience as starting points. Greek philosophy had recourse to the common forms of language to suggest its generalizations. It found the typical statement, 'That stone is grey'; and it evolved the generalization that the actual world can be conceived as a collection of primary substances qualified by universal qualities. Of course, this was not the only generalization evolved: Greek philosophy was subtle and multiform, also it was not inflexibly consistent. But this general notion was always influencing thought, explicitly or implicitly.

A theory of knowledge was also needed. Again philosophy started on a sound principle, that all knowledge is grounded on perception. Perception was then analysed, and found to be the awareness that a universal quality is qualifying a particular substance. Thus perception is the catching of a universal quality in the act of qualifying a particular substance. It was then asked, how the perceiver perceives; and the answer is by his organs of sensation. Thus the universal qualities which qualify the perceived substances are, in respect to the perceiver, his private sensations referred to particular substances other than himself. So far, the tradition of philosophy includes, among other elements, a factor of extreme objectivism in metaphysics, whereby the subject-predicate form of proposition is taken as expressing a fundamental metaphysical truth.

Descartes modified traditional philosophy in two opposite ways. He increased the metaphysical emphasis on the substance-quality forms of thought. The actual things 'required nothing but themselves in order to exist,' and were to be thought of in terms of their qualities, some of them essential attributes, and others accidental modes. He also laid down the principle, that those substances which are the subjects enjoying conscious experiences, provide the primary data for philosophy, namely, themselves as in the enjoyment of such experience. This is the famous subjectivist bias which entered into modern philosophy through Descartes. In this doctrine Descartes undoubtedly made the greatest philosophical discovery since the age of Plato and Aristotle. For his doctrine directly traversed the notion that the proposition, 'This stone is grey,' expresses a primary form of known fact from which metaphysics can start its generalizations. If we are to go back to the subjective enjoyment of experience, the type of primary starting point [for Descartes] is 'my perception of this stone as grey.'

But like Columbus who never visited America, Descartes missed the full sweep of his own discovery, and he and his successors, Locke and Hume, continued to construe the functionings of the subjective enjoyment of experience according to the substance-quality categories. Yet if the enjoyment of experience be the constitutive subjective fact, these categories have lost all claim to any fundamental character in metaphysics. Hume—to proceed at once to the consistent exponent of the method—looked for a universal quality to function as qualifying the mind, by way of explanation of its perceptive enjoyment. Now if we scan 'my perception of this stone as grey' in order to find a universal, the only available candidate is 'greyness.' Accordingly for Hume, 'greyness,' functioning as a sensation qualifying the mind, is a fundamental type of fact for metaphysical generalization. The result is Hume's simple impressions of sensation, which form the starting point of his philosophy. But this is an entire muddle, for the perceiving mind is not grey, and so grey is now made to perform a new rôle. From the original fact 'my perception of this stone as grey,' Hume extracts 'Awareness of sensation of greyness'; and puts it forward as the ultimate datum in this element of experience.

He has discarded the objective actuality of the stone-image in his search for a universal quality: this 'objective actuality' is Descartes' *'realitas objectiva.'* Hume's search was undertaken in obedience to a metaphysical principle which had lost all claim to validity, if the Cartesian discovery be accepted. He is then content with 'sensation of greyness,' which is just as much a particular as the original stone-image. He is aware of *'this* sensation of greyness.' What he has done is to assert arbitrarily the 'subjectivist' and 'sensationalist' principles as applying to the datum for experience: the notion *'this* sensation of greyness' has no reference to any other actual entity. Hume thus applies to the experiencing subject Descartes' principle, that it requires no other actual entity in order to exist. The fact that finally Hume criticizes the Cartesian notion of mind, does not alter the other fact that his antecedent arguments presuppose that notion.

It is to be noticed that Hume can only analyse the sensation in terms of an universal and of its realization in the prehending mind. For example, to take the first examples which in his *Treatise* he gives of such analysis, we find 'red,' 'scarlet,' 'orange,' 'sweet,' 'bitter.' Thus Hume describes 'impressions of sensation' in the exact terms in which the philosophy of organism describes conceptual feelings. They are the particular feelings of universals, and are not feelings of other particular existents exemplifying universals. Hume admits this identification, and can find no distinction except in 'force and vivacity.' He writes: "The first circumstance that strikes my eye, is the great resemblance between our impressions and ideas in every particular except their degree of force and vivacity."

Hume, in effect, agrees that 'mind' is a process of concrescence arising from primary data. In his account, these data are 'impressions of sensation'; and in such impressions no elements other than universals are discoverable. For the philosophy of organism, the primary data are always actual entities absorbed into feeling in virtue of certain universals shared alike by the objectified actuality and the experient subject. Descartes takes an intermediate position. He explains perception in Humian terms, but adds an apprehension of particular actual entities in virtue of an '*inspectio*' and a '*judicium*' effected by the mind (Meditations II and III). Here he is paving the way for Kant, and for the degradation of the world into 'mere appearance.'

All modern philosophy hinges round the difficulty of describing the world in terms of subject and predicate, substance and quality, particular and universal. The result always does violence to that immediate experience which we express in our actions, our hopes, our sympathies, our purposes, and which we enjoy in spite of our lack of phrases for its verbal analysis. We find ourselves in a buzzing[3] world, amid a democracy of fellow creatures; whereas, under some disguise or other, orthodox philosophy can only introduce us to solitary substances, each enjoying an illusory experience: "O Bottom, thou art changed! what do I see on thee?" The endeavour to interpret experience in accordance with the overpowering deliverance of common sense must bring us back to some restatement of Platonic realism, modified so as to avoid the pitfalls which the philosophical investigations of the seventeenth and eighteenth centuries have disclosed. We perceive other things which are in the world of actualities in the same sense as we are. Also our emotions are directed towards other things, including of course our bodily organs. These are our primary beliefs which philosophers proceed to dissect.

The true point of divergence is the false notion suggested by the contrast between the natural meanings of the words 'particular' and 'universal.' The 'particular' is thus conceived as being just its individual self with no necessary relevance to any other particular. It answers to Descartes' definition of substance: "And when we conceive of substance, we merely conceive an existent thing which requires nothing but itself in order to exist." (Cf. *Principles of Philosophy,* Part I.) This definition is a true derivative from Aristotle's definition: A primary substance is "neither asserted of a subject nor present in a subject." (Cf. *Aristotle* by W. D. Ross, Ch. II.)

The philosophy of organism fully accepts Descartes' discovery that subjective experiencing is the primary metaphysical situation

---

3 This epithet is, of course, borrowed from William James. [Footnote is Whitehead's.]

which is presented to metaphysics for analysis. This doctrine is the 'reformed subjectivist principle.' The reformed subjectivist principle adopted by the philosophy of organism is merely an alternative statement of the principle of relativity.[4] This principle states that it belongs to the nature of a 'being' that it is a potential for every 'becoming.' Thus all things are to be conceived as qualifications of actual occasions. The subjectivist principle is that the whole universe consists of elements disclosed in the analysis of the experiences of subjects. Process is the becoming of experience.

The principle of universal relativity directly traverses Aristotle's dictum, '(A substance) is not present in a subject.' On the contrary, according to this principle an actual entity *is* present in other actual entities. In fact, if we allow for degrees of relevance, and for negligible relevance, we must say that every actual entity is present in every other actual entity; Descartes' discovery on the side of subjectivism requires balancing by an 'objectivist' principle as to the datum for experience. Of course, strictly speaking, the universals, to which Hume confines the datum, are also 'objects'; but the phrase 'objective content' is meant to emphasize the doctrine of 'objectification' of actual entities. If experience be not based upon an objective content, there can be no escape from a solipsist subjectivism. Hume fails to provide experience with any objective content.

With the advent of Cartesian subjectivism, the substance-quality category has lost all claim to metaphysical primacy; and, with this deposition of substance-quality, we can reject the notion of individual substances, each with its private world of qualities and sensations. The philosophy of organism entirely accepts the subjectivist bias of modern philosophy. It also accepts Hume's doctrine that nothing is to be received into the philosophical scheme which is not discoverable as an element in subjective experience. This is the ontological principle. Thus Hume's demand that causation be describable as an element in experience is, on these principles, entirely justifiable. The point of the criticisms of Hume's procedure is that we have direct intuition of inheritance and memory: thus the only problem is, so to describe the general character of

4 See Section I of Chapter Two.

experience that these intuitions may be included. It is here that
Hume fails.

*The key point in the above is that "Descartes' discovery on the
side of subjectivism requires balancing by an 'objectivist' principle
as to the datum for experience." If one accepts Descartes' subjectivist bias, namely the view (which results from the* Cogito *argument) that subjects enjoying experiences provide the primary data
for philosophy, then accepting also the subjectivist principle—
namely, that all that this experience reveals is universals—lands
one inexorably in skepticism. Hume's virtue was to have seen this
clearly; Whitehead's rebuttal is to accept the subjectivist bias and
couple it with his "reformed" subjectivist principle, which states
that* actual entities, *and not merely* universals, *are revealed in experience. In this way the subjectivist bias is retained in philosophy
at the same time that the representative theory of perception is
repudiated.*

*The first sentence of the final paragraph of text above indicates
that "the advent of Cartesian subjectivism" means that "the substance-quality category has lost all claim to metaphysical primacy."
Whitehead now turns to a direct consideration of the substance-
quality category.*

Descartes asked the fundamental metaphysical question, What
is it to be an actual entity? He found three kinds of actual entities,
namely, cogitating minds, extended bodies, and God. His word
for an actual entity was 'substance.' The fundamental proposition,
whereby the analysis of actuality could be achieved, took the form
of predicating a quality of the substance in question. A quality
was either an accident or an essential attribute. In the Cartesian
philosophy there was room for three distinct kinds of change: one
was the change of accidents of an enduring substance; another was
the origination of an individual substance; and the third was the
cessation of the existence of an enduring substance. Any individual
belonging to either of the first two kinds of substances did not require any other individual of either of these kinds in order to
exist. But it did require the concurrence of God. Thus the essential attributes of a mind were its dependence on God and its

cogitations; and the essential attributes of a body were its dependence on God and its extension. Descartes does not apply the term
'attribute' to the 'dependence on God'; but it is an essential element in his philosophy.

It is quite obvious that the accidental relationships between diverse individual substances form a great difficulty for Descartes. If
they are to be included in his scheme of the actual world, they
must be qualities of a substance. Thus a relationship is the correlation of a pair of qualities one belonging exclusively to one individual, and the other exclusively to the other individual. The
correlation itself must be referred to God as one of his accidental
qualities. This is exactly Descartes' procedure in his theory of
representative ideas. In this theory, the perceived individual has
one quality; the perceiving individual has another quality which
is the 'idea' representing this quality; God is aware of the correlation; and the perceiver's knowledge of God guarantees for him
the veracity of his idea.

It is unnecessary to criticise this very artificial account of what
common sense believes to be our direct knowledge of other actual
entities. But it is the only account consistent with the metaphysical
materials provided by Descartes, combined with his assumption
of a multiplicity of actual entities. In this assumption of a multiplicity of actual entities the philosophy of organism follows Descartes. It is however obvious that there are only two ways out of
Descartes' difficulties; one way is to have recourse to some form
of monism; the other way is to reconstruct Descartes' metaphysical
machinery.

But Descartes asserts one principle which is the basis of all
philosophy: he holds that the whole pyramid of knowledge is based
upon the immediate operation of knowing which is either an essential (for Descartes), or a contributory, element in the composition of an immediate actual entity. This is also a first principle for
the philosophy of organism. But Descartes allowed the subject-
predicate form of proposition, and the philosophical tradition derived from it, to dictate his subsequent metaphysical development.
For his philosophy, 'actuality' meant 'to be a substance with inhering qualities.' For the philosophy of organism, the percipient

occasion is its own standard of actuality. If in its knowledge other actual entities appear, it can only be because they conform to its standard of actuality. There can only be evidence of a world of actual entities, if the immediate actual entity discloses them as essential to its own composition.

Descartes' notion of an unessential experience of the external world is entirely alien to the organic philosophy. This is the root point of divergence; and is the reason why the organic philosophy has to abandon any approach to the substance-quality notion of actuality. The organic philosophy interprets experience as meaning the 'self-enjoyment of being one among many, and of being one arising out of the composition of many.' Descartes interprets experience as meaning the 'self-enjoyment, by an individual substance, of its qualification by ideas.' The philosophies of substance presuppose a subject which then encounters a datum, and then reacts to the datum. The philosophy of organism presupposes a datum which is met with feelings, and progressively attains the unity of a subject. Descartes in his own philosophy conceives the thinker as creating the occasional thought. The philosophy of organism inverts the order, and conceives the thought as a constituent operation in the creation of the occasional thinker. The thinker is the final end whereby there is the thought. In this inversion we have the final contrast between a philosophy of substance and a philosophy of organism. The operations of an organism are directed towards the organism as a 'superject,' and are not directed from the organism as a 'subject.' The operations are directed *from* antecedent organisms and *to* the immediate organism. They are 'vectors,' in that they convey the many things into the constitution of the single superject. An actual entity is at once the product of the efficient past, and is also, in Spinoza's phrase, *causa sui*. Every philosophy recognizes, in some form or other, this factor of self-causation, in what it takes to be ultimate actual fact.

In the paragraphs that follow, Whitehead approaches the same problem (the problem involved when a philosophy admits independent realities—i.e., substances) in terms of its origin in the subject-predicate mode of thought derived from Aristotle.

The philosophy of organism is governed by the belief that the subject-predicate form of proposition is concerned with high abstractions, except in its application to subjective forms. This sort of abstraction, apart from this exception, is rarely relevant to metaphysical description. The dominance of Aristotelian logic from the late classical period onwards has imposed on metaphysical thought the categories naturally derivate from its phraseology. This dominance of his logic does not seem to have been characteristic of Aristotle's own metaphysical speculations. The divergencies, such as they are, in these lectures from other philosophical doctrines mostly depend upon the fact that many philosophers, who in their explicit statements criticize the Aristotelian notion of 'substance,' yet implicitly throughout their discussions presuppose that the 'subject-predicate' form of proposition embodies the finally adequate mode of statement about the actual world. The evil produced by the Aristotelian 'primary substance' is exactly this habit of metaphysical emphasis upon the 'subject-predicate' form of proposition.

The doctrine of the individual independence of real facts is derived from the notion that the subject-predicate form of statement conveys a truth which is metaphysically ultimate. According to this view, an individual substance with its predicates constitutes the ultimate type of actuality. If there be one individual, the philosophy is monistic; if there be many individuals, the philosophy is pluralistic. With this metaphysical presupposition, the relations between individual substances constitute metaphysical nuisances: there is no place for them. Accordingly—in defiance of the most obvious deliverance of our intuitive 'prejudices'—every respectable philosophy of the subject-predicate type is monistic.

The exclusive dominance of the substance-quality metaphysics was enormously promoted by the logical bias of the mediaeval period. It was retarded by the study of Plato and of Aristotle. These authors included the strains of thought which issued in this doctrine, but included them inconsistently mingled with other notions. The substance-quality metaphysics triumphed with exclusive dominance in Descartes' doctrines. Unfortunately he did not realize that his notion of the 'res vera' did not entail the same disjunction of ultimate facts as that entailed by the Aristotelian notion of 'primary substance.' Locke led a revolt from this dominance, but inconsistently. For him and also for Hume, in the background and tacitly presupposed in all explanations, there remained the mind with its perceptions. The perceptions, for Hume, are what the mind knows about itself; and tacitly the knowable facts are always treated as qualities of a subject—the subject being the mind. His final criticism of the notion of the 'mind' does not alter the plain fact that the whole of the previous discussion has included this presupposition. Hume's final criticism only exposes the metaphysical superficiality of his preceding exposition.

The doctrine of the "individual independence of real facts" raises additional problems when one explicitly recognizes that it implies the "individual independence of successive temporal occasions." Whitehead now turns his attention to this aspect of the problem.

We have certainly to make room in our philosophy for the two contrasted notions, one that every actual entity endures, and the other that every morning is a new fact with its measure of change. These various aspects can be summed up in the statement that *experience* involves a *becoming,* that *becoming* means that *something becomes,* and that *what becomes* involves *repetition* transformed into *novel immediacy.*

This statement directly traverses one main presupposition which Descartes and Hume agree in stating explicitly. This presupposition is that of the individual independence of successive temporal occasions. For example, Descartes writes: "(The nature of time) is such that its parts do not depend one upon the other. . . ." Also Hume's impressions are self-contained, and he can find no temporal relationship other than mere serial order. This statement about Hume requires qualifying so far as concerns the connection between 'impressions' and 'ideas.' There is a relation of 'derivation' of 'ideas' from 'impressions' which he is always citing and never discussing. So far as it is to be taken seriously—for he never refers it to a correlate 'impression'—it constitutes an exception to the individual independence of successive 'perceptions.' This presupposition of individual independence is what I have elsewhere (cf. *Science and the Modern World,* Ch. III) called, the 'fallacy of simple location.'

The notion of 'simple location' is inconsistent with any admission of 'repetition'; Hume's difficulties arise from the fact that he starts with simple locations and ends with repetition. In the organic philosophy the notion of repetition is fundamental. The doctrine of objectification is an endeavor to express how what is settled in actuality is repeated under limitations, so as to be 'given' for immediacy. In discussing 'time,' this doctrine will be termed the doctrine of 'objective immortality.'[5]

II

Locke (and Hume)

Locke explicitly discards metaphysics. His enquiry has a limited scope: "This therefore being my purpose, to inquire into the original, certainty and extent of human knowledge, together with the grounds and degrees of belief, opinion, and assent, I shall not at present meddle with the physical consideration of the mind, or trouble myself to examine wherein its essence consists, . . . It shall suffice to my present purpose, to consider the discerning faculties of a man as they are employed about the objects which they have to do with; . . ." (Cf. *Essay*, I, I, 2.) The enduring importance of Locke's work comes from the candour, clarity, and adequacy with which he stated the evidence, uninfluenced by the bias of metaphysical theory. He explained, in the sense of stating plainly, and not in the more usual sense of 'explaining away.' By an ironic development in the history of thought, Locke's successors, who arrogated to themselves the title of 'empiricists,' have been chiefly employed in explaining away the obvious facts of experience in obedience to the a priori doctrine of sensationalism, inherited from the mediaeval philosophy which they despised. Locke's *Essay* is the invaluable storehouse for those who wish to confront their metaphysical constructions by a recourse to the facts.

In Hume's philosophy the primary impressions are characterized in terms of universals, e.g. in the first section of his *Treatise* he refers to the colour 'red' as an illustration. This is also the doctrine

[5] Throughout the last part of this section Whitehead has been attacking the "substance-quality concept." He returns to the attack from the standpoint of modern science in the first part of Section VI.

of the first two books of Locke's *Essay*. But in Locke's third book a different doctrine appears, and the primary data are explicitly said to be 'ideas of particular existents.' According to Locke's second doctrine, the ideas of universals are derived from these primary data by a process of comparison and analysis. The philosophy of organism agrees in principle with this second doctrine of Locke's.

Hume clipped his explanation by this a priori theory, which he states explicitly in his *Treatise:* "We may observe, that it is universally allowed by philosophers, and is besides pretty obvious of itself, that nothing is ever really present with the mind but its perceptions or impressions and ideas, and that external objects become known to us only by those perceptions they occasion. To hate, to love, to think, to feel, to see; all this is nothing but to perceive."

Hume, in agreement with what 'is universally allowed by philosophers,' interprets this statement in a sensationalist sense. In accordance with this sense, an impression is nothing else than a particular instance of the mind's awareness of a universal, which may either be simple, or may be a manner of union of many simple universals. For Hume, hating, loving, thinking, feeling, are nothing but perceptions derivate from these fundamental impressions. This is the a priori sensationalist dogma, which bounds all Hume's discoveries in the realm of experience. It is probable that this dogma was in Locke's mind throughout the earlier portion of his *Essay*. But Locke was not seeking consistency with any a priori dogma. He also finds in experience 'ideas' with characteristics which 'determine them to this or that particular existent.' Such inconsistency with their dogma shocks empiricists, who refuse to admit experience, naked and unashamed, devoid of their a priori figleaf. Locke is merely stating what, in practice, nobody doubts.

The metaphysical superiority of Locke over Hume is exhibited in his wide use of the term 'idea,' which Locke himself introduced and Hume abandoned. Its use marks the fact that his tacit subject-predicate bias is slight in its warping effect. He first (I, I, 8) explains: ". . . I have used it (i.e. idea) to express whatever is meant by phantasm, notion, species, or whatever it is which the mind can be employed about in thinking. . . ." But later (III, III, 6 and 7), without any explicit notice of the widening of use, he writes: ". . . and ideas *become general* by separating from them the circumstances of time, and place, and any other ideas that may *determine them to this or that particular existence.*" (Italics mine.) Here, for Locke, the operations of the mind originate from ideas 'determined' to particular existents. This is a fundamental principle with Locke; it is a casual concession to the habits of language with Hume; and it is a fundamental principle with the philosophy of organism.

But Locke wavers in his use of this principle of some sort of perception of 'particular existents'; and Hume seeks consistency by abandoning it; while the philosophy of organism seeks to reconstruct Locke by abandoning those parts of his philosophy which are inconsistent with this principle. If Locke's *Essay* is to be interpreted as a consistent scheme of thought, undoubtedly Hume is right; but such an interpretation offers violence to Locke's contribution to philosophy. Spinoza is practically a logical systematization of Descartes, purging him of inconsistencies. But this attainment of logical coherence is obtained by emphasizing just those elements in Descartes which the philosophy of organism rejects. In this respect, Spinoza performs the same office for Descartes that Hume does for Locke. The philosophy of organism may be conceived as a recurrence to Descartes and to Locke, in respect to just those elements in their philosophies which are usually rejected by reason of their inconsistency with the elements which their successors developed. Thus the philosophy of organism is pluralistic in contrast with Spinoza's monism; and is a doctrine of experience prehending actualities, in contrast with Hume's sensationalist phenomenalism.

Hume has only impressions of 'sensation' and of 'reflection.' He writes: "The first kind arises in the soul originally, from unknown causes." (*Treatise,* Bk. I, Sect. II.) Note the tacit presupposition of 'the soul' as subject, and 'impression of sensation' as predicate. Also note the dismissal of any intrinsic relevance to a particular existent, which is an existent in the same sense as the 'soul' is an existent; whereas Locke illustrates his meaning by referring (cf. III, III, 7) to a 'child'—corresponding to 'the soul' in Hume's phrase—and to its 'nurse' *of whom* the child has its 'idea.'

Hume is certainly inconsistent, because he cannot entirely disregard common sense. But his inconsistencies are violent, and his main argument negates Locke's use. As an example of his glaring inconsistency of phraseology, note: "As to those *impressions,* which arise from the senses, their ultimate cause is, in my opinion, perfectly inexplicable by human reason, and it will always be impossible to decide with certainty, whether they arrive immediately from the object, or are produced by the creative power of the mind, or are derived from the Author of our being." (Cf. *Treatise,* Bk. III, Sects. V and VI.) Here he inconsistently speaks of *the object,* whereas he has nothing on hand in his philosophy which justifies the demonstrative word '*the.*' In the second reference '*the object*' has emerged into daylight. He writes: "There is no object which implies the existence of any other, if we consider these objects in themselves, and never look beyond the ideas which we form of them." This quotation exhibits an ingenious confusion whereby Hume makes the best of two metaphysical worlds, the world with Locke's principle, and his own world which is without Locke's principle.

But Locke's principle amounts to this: That there are many actual existents, and that in some sense one actual existent repeats itself in another actual existent, so that in the analysis of the latter existent a component 'determined to' the former existent is discoverable. The philosophy of organism expresses this principle by its doctrines of 'prehension' and of 'objectification.' Locke always supposes that consciousness is consciousness of the ideas in the conscious mind. But he never separates the 'ideas' from the 'consciousness.' The philosophy of organism makes this separation, and thereby relegates consciousness to a subordinate metaphysical position; and gives to Locke's *Essay* a metaphysical interpretation which was not in Locke's mind.

Locke has in fact stated in his work one main problem for the philosophy of organism. He discovers that the mind is a unity arising out of the active prehension of ideas into one concrete thing. Unfortunately, he presupposes both the Cartesian dualism whereby minds are one kind of particulars, and natural entities are another kind of particulars, and also the subject-predicate dogma. He is thus, in company with Descartes, driven to a theory of representative perception. For example, in reference to children he writes: "The ideas of the nurse and the mother are well framed in their minds; and, like pictures of them there, represent only those individuals" [III, III, 7].

[Descartes,] in his efforts to guard his representative 'ideas' from the fatal gap between mental symbol and actuality symbolized, practically, in some sentences, expresses the doctrine of objectification here put forward. Thus: "Hence the idea of the sun will be the sun itself existing in the mind, not indeed formally, as it exists in the sky, but objectively, i.e. in the way in which objects are wont to exist in the mind; and this mode of being is truly much less perfect than that in which things exist outside the mind, but it is not on that account mere nothing, as I have already said." (Cf. *Reply to Objections I.*)

Both Descartes and Locke, in order to close the gap between idea representing and 'actual entity represented,' require this doctrine of 'the sun itself existing in the mind.' But though, as in this passage, they at times casually state it in order to push aside the epistemological difficulty, they neither of them live up to these admissions. They relapse into the tacit presupposition of the mind with its private ideas which are in fact qualities without intelligible connection with the entities represented. This doctrine obviously creates an insoluble problem for epistemology, only to be solved either by some sturdy make-believe of 'animal faith,' with Santayana, or by some doctrine of illusoriness—some doctrine of mere appearance, inconsistent if taken as real—with Bradley. Anyhow 'representative perception' can never, within its own metaphysical doctrines, produce the title deeds to guarantee the validity of the representation of fact by idea.

Locke's nearest approach to the philosophy of organism, and—from the point of view of that doctrine—his main oversight, are best exemplified by the first section of his chapter, 'Of our complex Ideas of Substances' (II, XXIII, 1). He writes: "The mind, being, as I have declared, furnished with a great number of the simple ideas conveyed in by the senses, as they are found in exterior things, or by reflection on its own operations, takes notice, also, that a certain number of these simple ideas go constantly together; which being presumed to belong to one thing, and words being suited to common apprehensions, and made use of for quick dispatch, are called, so united in one subject, by one name; which, by inadvertency, we are apt afterward to talk of and consider as one simple idea, which indeed is a complication of many ideas together: because, as I have said, not imagining how these simple ideas can subsist by themselves, we accustom ourselves to suppose some *substratum* wherein they do subsist, and from which they do result; which therefore we call substance."

In this section, Locke's first statement, which is the basis of the remainder of the section, is exactly the primary assumption of the philosophy of organism: "The mind, being furnished with a great number of the simple ideas conveyed in by the senses, *as they are found in exterior things, . . .*" Here the last phrase, 'as they are found in exterior things,' asserted what I call the *vector* character of the primary feelings. The universals involved obtain that status by reason of the fact that '*they are found in exterior things.*' This is Locke's assertion and it is the assertion of the philosophy of organism. It can also be conceived as a development of Descartes' doctrine of '*realitas objectiva.*' The universals are the only elements in the data describable by concepts, because concepts are merely the analytic functioning of universals. But the 'exterior things,' although they are not expressible by concepts in respect to their individual particularity, are no less data for feeling; so that the concrescent actuality arises from feeling their status of individual particularity; and thus that particularity is included as an element from which feelings originate, and which they concern.

The sentence later proceeds with, "a certain number of these simple ideas go constantly together." This can only mean that in the immediate perception 'a certain number of these simple ideas' are found together in an exterior thing, and that the recollection of antecedent moments of experience discloses that the same fact, of togetherness in an exterior thing, holds for the same set of simple ideas. Again, the philosophy of organism agrees that this description is true for moments of immediate experience. But Locke, owing to the fact that he veils his second premise under the phrase 'go constantly together,' omits to consider the question whether the 'exterior things' of the successive moments are to be identified.

The answer of the philosophy of organism is that, in the sense in which Locke is here speaking, the exterior things of successive moments are not to be identified with each other. Each exterior thing is either one actual entity, or (more frequently) is a nexus of actual entities with immediacies mutually contemporary. For the sake of simplicity we will speak only of the simpler case where the 'exterior thing' means one actual entity at the moment in question. But what Locke is explicitly concerned with is the notion of the self-identity of the one enduring physical body which lasts for years, or for seconds, or for ages. He is considering the current philosophical notion of an individualized particular substance (in the Aristotelian sense) which undergoes adventures of change, retaining its substantial form amid transition of accidents. Throughout his *Essay*, he in effect retains this notion while rightly insisting on its vagueness and obscurity. The philosophy of organism agrees with Locke and Hume, that the non-individualized substantial form is nothing else than the collection of universals—or, more accurately, the one complex universal—common to the succession of 'exterior things' at successive moments respectively. In other words, an 'exterior thing' is either one 'actual entity,' or is a 'society' with a 'defining characteristic.' For the organic philosophy, these 'exterior things' (in the former sense) are the final concrete actualities. The individualized substance (of Locke) must be construed to be the historic route constituted by some society of fundamental 'exterior things,' stretching from the first 'thing' to the last 'thing.'

But Locke, throughout his *Essay*, rightly insists that the chief ingredient in the notion of 'substance' is the notion of 'power.' The philosophy of organism holds that in order to understand 'power,' we must have a correct notion of how each individual actual entity contributes to the datum *from which* its successors arise and *to which* they must conform. The reason why the doctrine of power is peculiarly relevant to the enduring things, which the philosophy of Locke's day conceived as individualized substances, is that any likeness between the successive occasions of an historic route procures a corresponding identity between their contributions to the datum of any subsequent actual entity; and it therefore secures a corresponding intensification in the imposition of conformity.

The perceptive constitution of the actual entity presents the problem, How can the other actual entities, each with its own formal existence, also enter objectively into the perceptive constitution of the actual entity in question? This is the problem of the solidarity of the universe. The classical doctrines of universals and particulars, of subject and predicate, of individual substances not present in other individual substances, of the externality of relations, alike render this problem incapable of solution. The answer given by the organic philosophy is the doctrine of prehensions, involved in concrescent integrations, and terminating in a definite, complex unity of feeling. To be actual must mean that all actual things are alike objects, enjoying objective immortality in fashioning creative actions; and that all actual things are subjects, each prehending the universe from which it arises.

III

Kant

In the following paragraphs Whitehead brings out clearly the essential opposition between his own position and Kant's. The crux of the difference is embodied in the following phrase: "For Kant, the world emerges from the subject; for the philosophy of organism, the subject emerges from the world."

In any metaphysical scheme founded upon the Kantian or Hegelian traditions, experience is the product of operations which lie among the higher of the human modes of functioning. For such schemes, ordered experience is the result of schematization of modes of *thought,* concerning causation, substance, quality, quantity.

The process by which experiential unity is attained, is thereby conceived in the guise of modes of thought. The exception is to be found in Kant's preliminary sections on 'Transcendental Aesthetic,' by which he provides space and time. But Kant, following Hume, assumes the radical disconnection of impressions *qua data;* and therefore conceives his transcendental aesthetic to be the mere description of a subjective process appropriating the data by orderliness of feeling.

The philosophy of organism aspires to construct a critique of pure feeling, in the philosophical position in which Kant put his *Critique of Pure Reason*. This should also supersede the remaining *Critiques* required in the Kantian philosophy. Thus in the organic philosophy Kant's 'Transcendental Aesthetic' becomes a distorted fragment of what should have been his main topic.[6] The datum includes its own interconnections, and the first stage of the process of feeling is the reception into the responsive conformity of feeling whereby the datum, which is mere potentiality, becomes the individualized basis for a complex unity of realization.

The philosophy of organism is the inversion of Kant's philosophy. *The Critique of Pure Reason* describes the process by which subjective data pass into the appearance of an objective world. The philosophy of organism seeks to describe how objective data pass into subjective satisfaction, and how order in the objective data provides intensity in the subjective satisfaction. For Kant, the world emerges from the subject; for the philosophy of organism, the subject emerges from the world—a 'superject' rather than a 'subject.' The word 'object' thus means an entity which is a potentiality for being a component in feeling; and the word 'subject' means the entity constituted by the process of feeling, and including this process. The feeler is the unity emergent from its own feelings; and feelings are the details of the process intermediary between this unity and its many data. The data are the potentials for feeling; that is to say, they are objects. The process is the elimination of indeterminateness of feeling from the unity of one subjective experience. The degree of order in the datum is measured by the degree of richness in the objective lure. The 'intensity' achieved belongs to the subjective form of the satisfaction.

It is useful to compare this analysis of the construction of an act of experience with Kant's. In the first place Kant's act of experience is essentially knowledge. Thus whatever is not knowledge is necessarily inchoate, and merely on its way to knowledge. In comparing Kant's procedure with that of the philosophy of organism, it must be remembered that an 'apparent' objective content is the end of Kant's process, and thus takes the place of 'satisfaction' in the process as analysed in the philosophy of organism. In Kant's phrase-

[6] See the editor's first commentary in Section I of this chapter.

ology at the beginning of *The Critique of Pure Reason,* this 'apparent' objective content is referred to as 'objects.' He also accepts Hume's sensationalist account of the datum. Kant places this sentence at the commencement of the *Critique:* "Objects therefore are given to us through our sensibility. Sensibility alone supplies us with intuitions. These intuitions become thought through the understanding, and hence arise conceptions." (Translation is Max Müller's.) This is expanded later in a form which makes Kant's adhesion to Hume's doctrine of the datum more explicit: "And here we see that the impressions of the senses give the first impulse to the whole faculty of knowledge with respect to them, and thus produce experience which consists of two very heterogeneous elements, namely, matter for knowledge, derived from the senses (*eine Materie zur Erkenntniss aus den Sinnen*) and a certain form according to which it is arranged, derived from the internal source of pure intuition and pure thought, first brought into action by the former, and then producing concepts." (Cf. *Transcendental Analytic,* Ch. II, Sect. I [Max Müller].) Also: "Thoughts without content are empty, intuitions without concepts are blind." (Cf. *Transcendental Logic,* Introduction, Sect. I.) In this last statement the philosophy of organism is in agreement with Kant; but for a different reason. It is agreed that the functioning of concepts is an essential factor in knowledge, so that 'intuitions without concepts are blind.' But for Kant, apart from concepts there is nothing to know; since objects related in a knowable world are the product of conceptual functioning whereby categoreal form is introduced into the sense-datum, which otherwise is intuited in the form of a mere spatio-temporal flux of sensations. Knowledge requires that this mere flux be particularized by conceptual functioning, whereby the flux is understood as a nexus of 'objects.' Thus for Kant the process whereby there is experience is a process from subjectivity to apparent objectivity. The philosophy of organism inverts this analysis, and explains the process as proceeding from objectivity to subjectivity, namely, from the objectivity, whereby the external world is a datum, to the subjectivity, whereby there is one individual experience. Thus, according to the philosophy of organism, in every act of experience there are objects for knowledge; but, apart from the inclusion of intellectual functioning in that act of experience, there is no knowledge.

Kant's acceptance of the subjectivist principle—i.e., of the principle that the world is known as the result of higher, conscious modes of knowing—is the source of the disagreement between Kant and Whitehead. Note Whitehead's opening sentence in the section just quoted, in which he states that for the Kantian-Hegelian traditions "experience is the product of operations which lie among the higher of the human modes of functioning." Whitehead's counterassertion, of course, is that, like Hume, Kant has ignored primitive experience in the mode of causal efficacy.

Traditional philosophy in its account of conscious perception has exclusively fixed attention on its pure conceptual side; and thereby has made difficulties for itself in the theory of knowledge. Locke, with his naïve good sense, assumes that perception involves more than this conceptual side; though he fails to grasp the inconsistency of this assumption with the extreme subjectivist sensationalist doctrine. Physical feelings form the non-conceptual element in our awareness of nature.

Locke upholds the direct conscious apprehension of 'things without' (*e.g. Essay*, II, XXI, 1), otherwise termed 'exterior things' (II, XXIII, 1), or 'this or that particular existence' (III, III, 6), and illustrated by an individual nurse and an individual mother (III, III, 7). In the philosophy of organism the nexus, which is the basis for such direct apprehension, is provided by the physical feelings. The philosophy of organism here takes the opposite road to that taken alike by Descartes and by Kant. Both of these philosophers accepted (Descartes with hesitations, and Kant without question) the traditional subjectivist sensationalism, and assigned the intuition of 'things without' peculiarly to the intelligence.

From the point of view of the philosophy of organism, the credit must be given to Hume that he emphasized the 'process' inherent in the fact of being a mind. His analysis of that process is faulty in its details. It was bound to be so; because, with Locke, he misconceived his problem to be the analysis of mental operations. He should have conceived it as the analysis of operations constituent of actual entities. He would then have found mental operations in their proper place.[7] Kant followed Hume in this misconception; and was thus led to balance the world upon thought—oblivious to the scanty supply of thinking.

IV

Newton

The purpose of this section is to throw light upon the rationale behind Whitehead's doctrine of the extensive continuum and his account of motion and change by examining his analysis of Newton's cosmology. This analysis presupposes Whitehead's discussion of the extensive continuum; the essence of this discussion can be obtained by reviewing Section II of Chapter Five.

The most general notions underlying the words 'space' and 'time' are those which this discussion [of the extensive continuum] has aimed at expressing in their true connection with the actual world. The alternative doctrine, which is the Newtonian cosmology, emphasized the 'receptacle' theory of space-time, and minimized the factor of potentiality. Thus bits of space and time were conceived as being as actual as anything else, and as being 'occupied' by other actualities which were the bits of matter. This is the Newtonian 'absolute' theory of space-time, which philosophers have never accepted, though at times some have acquiesced. Newton's famous *Scholium* (Andrew Motte's translation; new edition revised, London, 1803) to his first eight definitions in his *Principia* expresses this point of view with entire clearness:

"Hitherto I have laid down the definitions of such words as are less known, and explained the sense in which I would have them to be understood in the following discourse. I do not define time, space, place and motion, as being well known to all. Only I must observe, that the vulgar conceive those quantities under no other notions but from the relation they bear to sensible objects. And thence arise certain prejudices, for the removing of which, it will

---

[7] That is, as the culmination of experience and not as its foundation.

be convenient to distinguish them into absolute and relative, true and apparent, mathematical and common.

"I. Absolute, true, and mathematical time, of itself, and from its own nature, flows equably without regard to anything external, and by another name is called duration: relative, apparent, and common time, is some sensible and external (whether accurate or unequable) measure of duration by means of motion, which is commonly used instead of true time; such as an hour, a day, a month, a year.

"II. Absolute space, in its own nature, and without regard to anything external, remains always similar and immovable. Relative space is some movable dimension or measure of the absolute spaces; which our senses determine by its position to bodies, and which is vulgarly taken for immovable space; . . . absolute and relative space are the same in figure and magnitude; but they do not remain always numerically the same. . . .

"IV. . . . As the order of the parts of time is immutable, so also is the order of the parts of space. Suppose those parts to be moved out of their places, and they will be moved (if the expression may be allowed) out of themselves. For times and spaces are, as it were, the places as well of themselves as of all other things. All things are placed in time as to order of situation. It is from their essence or nature that they are places; and that the primary places of things should be movable, is absurd. These are, therefore, the absolute places; and translations out of those places are the only absolute motions. . . . Now no other places are immovable but those that, from infinity to infinity, do all retain the same given positions one to another; and upon this account must ever remain unmoved; and do thereby constitute, what I call, immovable space. The causes by which true and relative motions are distinguished, one from the other, are the forces impressed upon bodies to generate motion. True motion is neither generated nor altered, but by some force impressed upon the body moved: but relative motion may be generated or altered without any force impressed upon the body. For it is sufficient only to impress some force on other bodies with which the former is compared, that by their giving way, that relation may be changed, in which the relative rest or motion of this other body did consist. . . . The effects which distinguish absolute from relative motion are, the forces of receding from the axis of circular motion. For there are no such forces in a circular motion purely relative, but, in a true and absolute circular motion, they are greater or less, according to the quantity of motion. . . . Wherefore relative quantities are not the quantities themselves, whose names they bear, but those sensible measures of them (either accurate or inaccurate) which are commonly used instead of the measured quantities themselves. . . ."

I have quoted at such length from Newton's *Scholium* because this document constitutes the clearest, most definite, and most influential statement among the cosmological speculations of mankind, speculations of a type which first assume scientific importance with the Pythagorean school preceding and inspiring Plato. Newton is presupposing four types of entities which he does not discriminate in respect to their actuality: for him minds are actual things, bodies are actual things, absolute durations of time are actual things, and absolute places are actual things. He does not use the word 'actual'; but he is speaking of matter of fact, and he puts them all on the same level in that respect. The result is to land him in a clearly expressed but complex and arbitrary scheme of relationships between spaces *inter se;* between durations *inter se;* and between minds, bodies, times and places, for the conjunction of them all into the solidarity of the one universe. For the purposes of science it was an extraordinarily clarifying statement, that is to say, for all the purposes of science within the next two hundred years, and for most of its purposes since that period. But, as a fundamental statement, it lies completely open to sceptical attack; and also, as Newton himself admits, diverges from common sense—"the vulgar conceive those quantities under no other notions but from the relation they bear to sensible objects." Kant only saved it by reducing it to the description of a construct by means of which 'pure intuition' introduces an order for chaotic data; and for the schools of transcendentalists derived from Kant this construct has remained in the inferior position of a derivative from the proper ultimate substantial reality. For them it is an element in 'appearance'; and appearance is to be distinguished from reality.

The philosophy of organism is an attempt, with the minimum of critical adjustment, to return to the conceptions of the 'vulgar.' In the first place, the discussion must fasten on the notion of a 'sensible object,' to quote Newton's phrase. We may expand Newton's phrase, and state that the common sense of mankind conceives that all its notions ultimately refer to actual entities, or as Newton terms them, 'sensible objects.' Newton, basing himself upon current phys-ical notions, conceived 'sensible objects' to be the material bodies to which the science of dynamics applies. He was then left with the antithesis between 'sensible objects' and empty space. Newton, in-deed, as a private opinion, conjectured that there is a material medium pervading space. But he also held that there *might* not be such a medium. For him the notion 'empty space'—that is, mere spatiality—had sense, conceived as an independent actual existence 'from infinity to infinity.' In this he differed from Descartes. Modern physics sides with Descartes. It has introduced the notion of the 'physical field.' Also the latest speculations tend to remove the sharp distinction between the 'occupied' portions of the field and the 'unoccupied' portion.

According to Newton, a portion of space cannot move. We have to ask how this truth, obvious from Newton's point of view, takes shape in the organic theory.

The extensive continuum is that general relational element in experience whereby the actual entities experienced, and that unit experience itself, are united in the solidarity of one common world. The actual entities atomize it, and thereby make real what was antecedently merely potential. The atomization of the extensive continuum is also its temporalization; that is to say, it is the process of the becoming of actuality into what in itself is merely potential. The systematic scheme, in its completeness embracing the actual past and the potential future, is prehended in the positive experi-ence of each actual entity. In this sense, it is Kant's 'form of intui-tion'; but it is derived from the actual world *qua datum,* and thus is not 'pure' in Kant's sense of that term. It is not productive of the ordered world, but derivative from it.

Newton in his description of space and time has confused what is 'real' potentiality with what is actual fact. He has thereby been led to diverge from the judgment of 'the vulgar' who 'conceive those quantities under no other notions but from the relation they bear to sensible objects.' The philosophy of organism starts by agreeing with 'the vulgar' except that the term 'sensible object' is replaced by 'actual entity'; so as to free our notions from participation in an epistemological theory as to sense-perception. When we further consider how to adjust Newton's other descriptions to the organic theory, the surprising fact emerges that we must identify the atomized quantum of extension correlative to an actual entity, with Newton's absolute place and absolute duration. Newton's proof that motion does not apply to absolute place, which in its nature is immovable, also holds. Thus an actual entity never moves: it is where it is and what it is. Every actual entity is what it is, and is with its definite status in the universe, determined by its internal relations to other actual entities. The doctrine of internal relations makes it impossible to attribute 'change' to any actual entity.

An event is a nexus of actual occasions inter-related in some determinate fashion in some extensive quantum. For example, a molecule is a historic route of actual occasions; and such a route is an 'event.' The most general sense of the meaning of change is 'the differences between actual occasions in one event.' Now the motion of the molecule is nothing else than the differences between the successive occasions of its life-history in respect to the extensive quanta from which they arise; and the changes in the molecule are the consequential differences in the actual occasions. 'Change' is the description of the adventures of eternal objects in the evolving universe of actual things.

From the point of view of the philosophy of organism, there is great merit in Newton's immovable receptacles. But for Newton they are eternal. Locke's notion of time hits the mark better: time is 'perpetually perishing.' In the organic philosophy an actual entity has 'perished' when it is complete. Its birth is its end. The pragmatic use of the actual entity, constituting its static life, lies in the future. The creature perishes *and* is immortal. The actual entities beyond it can say, 'It is mine.' But the possession imposes conformation.

This conception of an actual entity in the fluent world is little more than an expansion of a sentence in the *Timaeus* (Jowett's translation): "But that which is conceived by opinion with the help of sensation and without reason, is always in the process of becoming and perishing and never really is." Bergson, in his protest against "spatialization," is only echoing Plato's phrase 'and never really is.'

V

Newton and Plato

In this section the discussion of Newton is continued, but in order to distinguish his own cosmology from that of Newton, Whitehead inserts an analysis of Plato's Timaeus *for the purpose of showing how the philosophy of organism appeals to Plato at precisely the points at which it departs from the Newtonian view.*

The *Timaeus* of Plato, and the *Scholium* of Newton—the latter already in large part quoted—are the two statements of cosmological theory which have had the chief influence on Western thought. In attempting an enterprise of the same kind, it is wise to follow the clue that perhaps the true solution consists in a fusion of the two previous schemes, with modifications demanded by self-consistency and the advance of knowledge. The cosmology explained in these lectures has been framed in accordance with this reliance on the positive value of the philosophical tradition.

To the modern reader, the *Timaeus,* considered as a statement of scientific details, is in comparison with the *Scholium* simply foolish. But what it lacks in superficial detail, it makes up for by its philosophic depth. If it be read as an allegory, it conveys profound truth; whereas the *Scholium* is an immensely able statement of details which, although abstract and inadequate as a philosophy, can within certain limits be thoroughly trusted for the deduction of truths at the same level of abstraction as itself. The penalty of its philosophical deficiency is that the *Scholium* conveys no hint of the limits of its own application. The practical effect is that the readers, and almost certainly Newton himself, so construe its meaning as to fall into what I have elsewhere (cf. *Science and the Modern World,* Ch. III) termed the 'fallacy of misplaced concreteness.'[8] It is the office of metaphysics to determine the limits of the applicability of such abstract notions.

8 See Sub-section 1 of Section VI for further illustration of this fallacy.

160 A Key to Whitehead's *Process and Reality*

The *Scholium* betrays its abstractness by affording no hint of that aspect of self-production, of generation, of φύσις, of *natura naturans,* which is so prominent in nature. For the *Scholium,* nature is merely, and completely, *there,* externally designed and obedient. The full sweep of the modern doctrine of evolution would have confused the Newton of the *Scholium,* but would have enlightened the Plato of the *Timaeus.* So far as Newton is concerned, we have his own word for this statement. In a letter to Bentley, he writes: "When I wrote my treatise about our system, I had an eye upon such principles as might work with considering men for the belief of a Deity; . . ." (This quotation is taken from Jebb's *Life of Bentley,* Ch. II.) The concept in Newton's mind is that of a fully articulated system requiring a definite supernatural origin with that articulation. This is the form of the cosmological argument, now generally abandoned as invalid; because our notion of causation concerns the relations of states of things within the actual world, and can only be illegitimately extended to a transcendent derivation. The notion of God, which will be discussed later,[9] is that of an actual entity immanent in the actual world, but transcending any finite cosmic epoch—a being at once actual, eternal, immanent, and transcendent. The transcendence of God is not peculiar to him. Every actual entity, in virtue of its novelty, transcends its universe, God included.

In the *Scholium,* space and time, with all their current mathematical properties, are ready-made for the material masses; the material masses are ready-made for the 'forces' which constitute their action and reaction; and space, and time, and material masses, and forces, are alike ready-made for the initial motions which the Deity impresses throughout the universe. It is not possible to extract from the *Scholium*—construed with misplaced concreteness— either a theism, or an atheism, or an epistemology, which can survive a comparison with the facts. This is the inescapable conclusion to be inferred from Hume's *Dialogues Concerning Natural Religion.* Biology is also reduced to a mystery; and finally physics itself has now reached a stage of experimental knowledge inexplicable in terms of the categories of the *Scholium.*

[9] See Chapter Seven.

In the *Timaeus*, there are many phrases and statements which find their final lucid expression in the *Scholium*. While noting this concurrence of the two great cosmological documents guiding Western thought, it cannot be too clearly understood that, within its limits of abstraction, what the *Scholium* says is true, and that it is expressed with the lucidity of genius. Thus any cosmological document which cannot be read as an interpretation of the *Scholium* is worthless. But there is another side to the *Timaeus* which finds no analogy in the *Scholium*. In general terms, this side of the *Timaeus* may be termed its metaphysical character, that is to say, its endeavour to connect the *behaviour* of things with the *formal nature* of things. The behaviour apart from the things is abstract, and so are the things apart from their behaviour. Newton —wisely, for his purposes—made this abstraction which the *Timaeus* endeavours to avoid.

In the first place, the *Timaeus* connects behaviour with the ultimate molecular characters of the actual entities. Plato conceives the notion of definite societies of actual molecular entities, each society with its defining characteristics. He does not conceive this assemblage of societies as *causa sui*. But he does conceive it as the work of subordinate deities, who are the animating principles of those departments of nature. In Greek thought, either poetic or philosophic, the separation between the φύσις and such deities had not that absolute character which it has for us who have inherited the Semitic Jehovah.

Newton could have accepted a molecular theory as easily as Plato, but there is this difference between them: Newton would have been surprised at the modern quantum theory and at the dissolution of quanta into vibrations; Plato would have expected it. While we note the many things said by Plato in the *Timaeus* which are now foolishness, we must also give him credit for that aspect of his teaching in which he was two thousand years ahead of his time. Plato accounted for the sharp-cut differences between kinds of natural things, by assuming an approximation of the molecules of the fundamental kinds respectively to the mathematical forms of the regular solids. He also assumed that certain qualitative contrasts in occurrences, such as that between musical notes, depended on the participation of these occurrences in some of the simpler ratios between integral numbers. He thus obtained a reason why there should be an approximation to sharp-cut differences between kinds of molecules, and why there should be sharp-cut relations of harmony standing out amid dissonance. Thus 'contrast'—as the opposite of incompatibility—depends on a certain simplicity of circumstance; but the higher contrasts depend on the assemblage of a multiplicity of lower contrasts, this assemblage again exhibiting higher types of simplicity.

It is well to remember that the modern quantum theory, with its surprises in dealing with the atom, is only the latest instance of a well-marked character of nature, which in each particular instance is only explained by some *ad hoc* dogmatic assumption. The theory of biological evolution would not in itself lead us to expect the sharply distinguished genera and species which we find in nature. There might be an occasional bunching of individuals round certain typical forms; but there is no explanation of the almost complete absence of intermediate forms. Again Newton's *Scholium* gives no hint of the ninety-two possibilities for atoms, or of the limited number of ways in which atoms can be combined so as to form molecules. Physicists are now explaining these chemical facts by means of conceptions which Plato would have welcomed.

There is another point in which the organic philosophy only repeats Plato. In the *Timaeus* the origin of the present cosmic epoch is traced back to an aboriginal disorder, chaotic according to our ideals. This is the evolutionary doctrine of the philosophy of organism. Plato's notion has puzzled critics who are obsessed with the Semitic theory of a wholly transcendent God creating out of nothing an accidental universe. Newton held the Semitic theory. The *Scholium* made no provision for the evolution of matter— very naturally, since the topic lay outside its scope. The result has been that the non-evolution of matter has been a tacit presupposition throughout modern thought. Until the last few years the sole alternatives were: either the material universe, with its present type of order, is eternal; or else it came into being, and will pass out of being, according to the fiat of Jehovah. Thus, on all sides,

Plato's allegory of the evolution of a new type of order based on new types of dominant societies became a daydream, puzzling to commentators.

Milton, curiously enough, in his *Paradise Lost* wavers between the *Timaeus* and the Semitic doctrine. This is only another instance of the intermixture of classical and Hebrew notions on which his charm of thought depends. In the description of Satan's journey across Chaos, Satan discovers

> The secrets of the hoary deep, a dark
> Illimitable ocean, without bound,
> Without dimension, where length, breadth and highth,
> And time and place are lost; where eldest Night
> And Chaos, ancestors of Nature, hold
> Eternal anarchy amidst the noise
> Of endless wars, and by confusion stand.

Milton is here performing for Plato the same poetic service that Lucretius performed for Democritus—with less justification, since Plato was quite capable of being his own poet. Also the fact of Satan's journey helped to evolve order; for he left a permanent track, useful for the devils and the damned.

The appeal to Plato in this section has been an appeal to the facts against the modes of expression prevalent in the last few centuries. These recent modes of expression are partly the outcome of a mixture of theology and philosophy, and are partly due to the Newtonian physics, no longer accepted as a fundamental statement. But language and thought have been framed according to that mould; and it is necessary to remind ourselves that this is not the way in which the world has been described by some of the greatest intellects. Both for Plato and for Aristotle the process of the actual world has been conceived as a real incoming of forms into real potentiality, issuing into that real togetherness which is an actual thing. Also, for the *Timaeus,* the creation of the world is the incoming of a type of order establishing a cosmic epoch. It is not the beginning of matter of fact, but the incoming of a certain type of social order.[10]

[10] Whitehead's discussion of levels of social order in Section II of Chapter Four provides the context for these remarks about the *Timaeus.*

VI

Modern Science

In many of the preceding passages Whitehead has registered protests against the notion of substance, against the concept of a "vacuous material existence" that endures passively through accidental adventures. Aristotelian logic, bedded in scientific materialism, has foisted this concept onto philosophy with disastrous results, Whitehead argues. In Section 1 of the material that follows he first explains why the concept of substance has come to dominate language and to haunt science and philosophy. He then argues that modern science forces one to the conclusion that the notion of substance, though still useful at the high level of abstraction involved in everyday discourse, is sheer error in metaphysics.

Following the discussion of substance, Whitehead turns in Section 2 to the more general question of the relationship of his philosophy of organism to contemporary science. He argues that modern philosophy in general has failed to throw light on scientific principles. By way of contrast he provides many illustrations to prove that "the general principles of physics are exactly what we should expect as a specific exemplification of the metaphysics required by the philosophy of organism."

1. Substance

The baseless metaphysical doctrine of 'undifferentiated endurance' is a subordinate derivative from the misapprehension of the proper character of the extensive scheme.

In our perception of the contemporary world via presentational immediacy, nexūs of actual entities are objectified for the percipient under the perspective of their characters of extensive continuity. In the perception of a contemporary stone, for example, the separate individuality of each actual entity in the nexus constituting the stone is merged into the unity of the extensive plenum, which for Descartes and for common sense, *is* the stone. The complete objectification is effected by the generic extensive perspective of the stone, specialized into the specific perspective of some sense-datum, such as some definite colour, for example. Thus the immediate percept assumes the character of the quiet undifferentiated endurance of the material stone, perceived by means of its quality of colour. This basic notion dominates language, and haunts both science and philosophy. Further, by an unfortunate application of the excellent maxim, that our conjectural explanation should always proceed by the utilization of a *vera causa,* whenever science or philosophy has ventured to extrapolate beyond the limits of the immediate deliverance of direct perception, a satisfactory explanation has always complied with the condition that substances with undifferentiated endurance of essential attributes be produced, and that activity be explained as the occasional modification of their accidental qualities and relations. Thus the imaginations of men are dominated by the quiet extensive stone with its relationships of positions, and its quality of colour—relationships and qualities which occasionally change. The stone, thus interpreted, guarantees the *vera causa,* and conjectural explanations in science and philosophy follow its model.

Thus in framing cosmological theory, the notion of continuous stuff with permanent attributes, enduring without differentiation, and retaining its self-identity through any stretch of time however small or large, has been fundamental. The stuff undergoes change in respect to accidental qualities and relations; but it is numerically self-identical in its character of one actual entity throughout its accidental adventures. The admission of this fundamental metaphysical concept has wrecked the various systems of pluralistic realism.

This metaphysical concept has formed the basis of scientific materialism. But the interpretation of the stone, on which the whole concept is based, has proved to be entirely mistaken. In the first place, from the seventeenth century onwards the notion of the simple inherence of the colour in the stone has had to be given up. This introduces the further difficulty that it is the colour which is extended and only inferentially the stone, since now we have had to separate the colour from the stone. Secondly, the molecular theory has robbed the stone of its continuity, of its unity, and of its passiveness. The stone is now conceived as a society of separate molecules in violent agitation. But the metaphysical concepts, which had their origin in a mistake about the stone, were now applied to the individual molecules. Each atom was still a stuff which retained its self-identity and its essential attributes in any portion of time—however short, and however long—provided that it did not perish. The notion of the undifferentiated endurance of substances with essential attributes and with accidental adventures was still applied. This is the root doctrine of materialism: the substance, thus conceived, is the ultimate actual entity.

But this materialistic concept has proved to be as mistaken for the atom as it was for the stone. The atom is only explicable as a society with activities involving rhythms with their definite periods. Again the concept shifted its application: protons and electrons were conceived as materialistic electric charges whose activities could be construed as locomotive adventures. We are now approaching the limits of any reasonable certainty in our scientific knowledge; but again there is evidence that the concept may be mistaken. The mysterious quanta of energy have made their appearance, derived, as it would seem, from the recesses of protons, or of electrons. Still worse for the concept, these quanta seem to dissolve into the vibrations of light. Also the material of the stars seems to be wasting itself in the production of the vibrations.

Further, the quanta of energy are associated by a simple law with the periodic rhythms which we detect in the molecules. Thus the quanta are, themselves, in their own nature, somehow vibratory; but they emanate from the protons and electrons. Thus there is every reason to believe that rhythmic periods cannot be dissociated from the protonic and electronic entities.

The same concept has been applied in other connections where it even more obviously fails. It is said that 'men are rational.' This is palpably false: they are only intermittently rational—merely liable to rationality. The intellect of Socrates is intermittent: he occasionally sleeps and he can be drugged or stunned.

The Cartesian subjectivism in its application to physical science became Newton's assumption of individually existent physical bodies, with merely external relationships. We diverge from Descartes by holding that what he has described as primary *attributes* of physical bodies, are really the forms of internal relationships *between* actual occasions, and *within* actual occasions. Such a change of thought is the shift from materialism to organism, as the basic idea of physical science.

In the language of physical science, the change from materialism to 'organic realism'—as the new outlook may be termed—is the displacement of the notion of static stuff by the notion of fluent energy. Such energy has its structure of action and flow, and is inconceivable apart from such structure. It is also conditioned by 'quantum' requirements. These are the reflections into physical science of the individual prehensions, and of the individual actual entities to which these prehensions belong. Mathematical physics translates the saying of Heraclitus, 'All things flow,' into its own language. It then becomes, All things are vectors. Mathematical physics also accepts the atomistic doctrine of Democritus. It translates it into the phrase, All flow of energy obeys 'quantum' conditions.

But what has vanished from the field of ultimate scientific conceptions is the notion of vacuous material existence with passive endurance, with primary individual attributes, and with accidental adventures. Some features of the physical world can be expressed in that way. But the concept is useless as an ultimate notion in science, and in cosmology.

The simple notion of an enduring substance sustaining persistent qualities, either essentially or accidentally, expresses an abstraction useful for many purposes of life. But whenever we try to use it as a fundamental statement of the nature of things, it proves itself mistaken. It arose from a mistake and has never succeeded in any of its applications. But it has had one success: it has entrenched itself in language, in Aristotelian logic, and in metaphysics. For its employment in language and in logic, there is—as stated above —a sound pragmatic defence. But in metaphysics the concept is sheer error. This error does not consist in the employment of the word 'substance'; but in the employment of the notion of an actual entity which is characterized by essential qualities, and remains numerically one amidst the changes of accidental relations and of accidental qualities. The contrary doctrine [the doctrine of these lectures] is that an actual entity never changes, and that it is the outcome of whatever can be ascribed to it in the way of quality or relationship.

2. Philosophy and Science

Whitehead now turns to the more general question of the relationship between his system and modern science. He first adumbrates his theory of prehension and then proceeds to show that "this metaphysical description of the simplest elements in the constitution of actual entities agrees absolutely with the general principles according to which the notions of modern physics are framed."

The experience of the simplest grade of actual entity is to be conceived as the unoriginative response to the datum with its simple content of sensa. The datum is simple, because it presents the objectified experiences of the past under the guise of simplicity. Occasions A, B, and C enter into the experience of occasion M as themselves experiencing sensa s_1 and s_2 unified by some faint contrast between s_1 and s_2. Occasion M responsively feels sensa s_1 and s_2 as its own sensations. There is thus a transmission of sensation emotion from A, B, and C to M.

Generalizing from the language of physics, the experience of M is an intensity arising out of specific sensa, directed from A, B, C. There is in fact a directed influx from A, B, C of quantitative feeling, arising from specific forms of feeling. The experience has a vector character, a common measure of intensity, and specific forms of feelings conveying that intensity. If we substitute the term 'energy' for the concept of a quantitative emotional intensity, and the term 'form of energy' for the concept of 'specific form of feeling,' and remember that in physics 'vector' means definite transmission from elsewhere, we see that this metaphysical description of the simplest elements in the constitution of actual entities agrees absolutely with the general principles according to which the notions of modern physics are framed. The 'datum' in metaphysics is the basis of the vector-theory in physics; the quantitative satisfaction in metaphysics is the basis of the scalar localization of energy in physics; the 'sensa' in metaphysics are the basis of the diversity of specific forms under which energy clothes itself. Scientific descriptions are, of course, entwined with the specific details of geometry and physical laws, which arise from the special order of the cosmic epoch in which we find ourselves. But the general principles of physics are exactly what we should expect as a specific exemplification of the metaphysics required by the philosophy of organism. It has been a defect in the modern philosophies that they throw no light whatever on any scientific principles. Science should investigate particular species, and metaphysics should investigate the generic notions under which those specific principles should fall. Yet, modern realisms have had nothing to say about scientific principles; and modern idealisms have merely contributed the unhelpful suggestion that the phenomenal world is one of the inferior avocations of the Absolute.

[In the philosophy of organism] an endeavour has been made to do justice alike to the aspect of the world emphasized by Descartes and to the atomism of the modern quantum theory. Descartes saw the natural world as an extensive spatial plenum, enduring through time. Modern physicists see energy transferred in definite quanta. This quantum theory also has analogues in recent neurology. Again fatigue is the expression of cumulation; it is physical memory. Further, causation and physical memory spring from the same root: both of them are physical perception. Cosmology must do equal justice to atomism, to continuity, to causation, to memory, to perception, to qualitative and quantitative forms of energy, and to extension. But so far [i.e., in philosophy up to Whitehead] there has been no reference to the ultimate vibratory characters of organisms and to the 'potential' element in nature.

The direct perception whereby the datum in the immediate subject is inherited from the past can thus, under an abstraction, be conceived as the transference of throbs of emotional energy, clothed in the specific forms provided by sensa. This direct perception, characterized by mere subjective responsiveness and by lack of origination in the higher phases, exhibits the constitution of an actual entity under the guise of receptivity. In the language of causation, it describes the efficient causation operative in the actual world. In the language of epistemology, as framed by Locke, it describes how the ideas of particular existents are absorbed into the subjectivity of the percipient and are the datum for its experience of the external world. In the language of science, it describes how the quantitative intensity of localized energy bears in itself the vector marks of its origin, and the specialities of its specific forms; it also gives a reason for the atomic quanta to be discerned in the building up of a quantity of energy. In this way, the philosophy of organism—as it should—appeals to the facts.

Chapter Seven

GOD AND THE WORLD

The subtitle of Process and Reality *is* An Essay in Cosmology. *This chapter presents the climactic sections of* Process and Reality, *the sections that draw into a* Weltanschauung *all the concepts and distinctions that have been presented in earlier chapters. Here the full cosmological sweep of Whitehead's metaphysical speculation emerges explicitly. Speaking of this portion of the book, he notes that it "is concerned with the final interpretation of the ultimate way in which the cosmological problem is to be conceived. It answers the question, What does it all come to?" This chapter is divided into two sections. The first, titled "The Ideal Opposites," sets the cosmological problem. The second, titled "God and the World," presents Whitehead's resolution.*

The two central pairs of ideal opposites that emerge in the first section (i.e., the two sets of opposites each pole of which is a locus of ideals) are (1) permanence and flux, and (2) order and novelty. Whitehead skillfully educes the profound significance of the paradoxes created by the legitimate ideals clustered around each pole and at the close of the section vividly establishes the problem to be resolved.

I

The Ideal Opposites

The chief danger to philosophy is narrowness in the selection of evidence. This narrowness arises from the idiosyncrasies and timidities of particular authors, of particular social groups, of particular schools of thought, of particular epochs in the history of civilization. The evidence relied upon is arbitrarily biased by the temperaments of individuals, by the provincialities of groups, and by the limitations of schemes of thought. The evil, resulting from this distortion of evidence, is at its worst in the consideration of the topic of the final part of this investigation—ultimate ideals. We must commence this topic by an endeavour to state impartially the general types of the great ideals which have prevailed at sundry seasons and places. Philosophy may not neglect the multifariousness of the world—the fairies dance, and Christ is nailed to the cross.

Ideals fashion themselves round two notions, permanence and flux. In the inescapable flux, there is something that abides; in the overwhelming permanence, there is an element that escapes into flux. Permanence can be snatched only out of flux; and the passing moment can find its adequate intensity only by its submission to permanence. Those who would disjoin the two elements can find no interpretation of patent facts.

That 'all things flow' is the first vague generalization which the unsystematized, barely analysed, intuition of men has produced. It is the theme of some of the best Hebrew poetry in the Psalms; it appears as one of the first generalizations of Greek philosophy in the form of the saying of Heraclitus; amid the later barbarism of Anglo-Saxon thought it reappears in the story of the sparrow flitting through the banqueting hall of the Northumbrian king; and in all stages of civilization its recollection lends its pathos to poetry. Without doubt, if we are to go back to that ultimate, integral experience, unwarped by the sophistications of theory, that experience whose elucidation is the final aim of philosophy, the flux of things is one ultimate generalization around which we must weave our philosophical system.

But there is a rival notion, antithetical to the former. I cannot at the moment recall one immortal phrase which expresses it with the same completeness as the alternative notion has been rendered by Heraclitus. This other notion dwells on permanences of things —the solid earth, the mountains, the stones, the Egyptian Pyramids, the spirit of man, God.

The best rendering of integral experience, expressing its general form divested of irrelevant details, is often to be found in the utterances of religious aspiration. One of the reasons of the thinness of so much modern metaphysics is its neglect of this wealth of expression of ultimate feeling. Accordingly we find in the first two lines of a famous hymn a full expression of the union of the two notions in one integral experience:

> Abide with me;
> Fast falls the eventide.

Here the first line expresses the permanences, 'abide,' 'me' and the 'Being' addressed; and the second line sets these permanences amid the inescapable flux. Here at length we find formulated the complete problem of metaphysics. Those philosophers who start with the first line have given us the metaphysics of 'substance'; and those who start with the second line have developed the metaphysics of 'flux.' But, in truth, the two lines cannot be torn apart in this way; and we find that a wavering balance between the two is a characteristic of the greater number of philosophers.

The four symbolic figures in the Medici chapel in Florence— Michelangelo's masterpieces of statuary, Day and Night, Evening and Dawn—exhibit the everlasting elements in the passage of fact. The figures stay there, reclining in their recurring sequence, forever showing the essences in the nature of things. The perfect realization is not merely the exemplification of what in abstraction is timeless. It does more: it implants timelessness on what in its essence is passing. The perfect moment is fadeless in the lapse of time. Time has then lost its character of 'perpetual perishing'; it becomes the 'moving image of eternity.'

Another contrast is equally essential for the understanding of ideals—the contrast between order as the condition for excellence, and order as stifling the freshness of living. This contrast is met with in the theory of education. The condition for excellence is a thorough training in technique. Sheer skill must pass out of the sphere of conscious exercise, and must have assumed the character of unconscious habit. The first, the second, and the third condition for high achievement is scholarship, in that enlarged sense including knowledge and acquired instinct controlling action.

The paradox which wrecks so many promising theories of education is that the training which produces skill is so very apt to stifle imaginative zest. Skill demands repetition, and imaginative zest is tinged with impulse. Up to a certain point each gain in skill opens new paths for the imagination. But in each individual formal training has its limit of usefulness. Beyond that limit there is degeneration: 'The lilies of the field toil not, neither do they spin.'

The social history of mankind exhibits great organizations in their alternating functions of conditions for progress, and of contrivances for stunting humanity. The history of the Mediterranean lands, and of western Europe, is the history of the blessing and the curse of political organizations, of religious organizations, of schemes of thought, of social agencies for large purposes. The moment of dominance, prayed for, worked for, sacrificed for, by generations of the noblest spirits, marks the turning point where the blessing passes into the curse. Some new principle of refreshment is required. The art of progress is to preserve order amid change, and to preserve change amid order. Life refuses to be embalmed alive. The more prolonged the halt in some unrelieved system of order, the greater the crash of the dead society.

The same principle is exhibited by the tedium arising from the unrelieved dominance of a fashion in art. Europe, having covered itself with treasures of Gothic architecture, entered upon generations of satiation. These jaded epochs seem to have lost all sense of that particular form of loveliness. It seems as though the last delicacies of feeling require some element of novelty to relieve their massive inheritance from bygone system. Order is not sufficient. What is required, is something much more complex. It is order entering upon novelty; so that the massiveness of order does not degenerate into mere repetition; and so that the novelty is always reflected upon a background of system.

But the two elements must not really be disjoined. It belongs to the goodness of the world, that its settled order should deal tenderly with the faint discordant light of the dawn of another age. Also order, as it sinks into the background before new conditions, has its requirements. The old dominance should be transformed into the firm foundations, upon which new feelings arise, drawing their intensities from delicacies of contrast between system and freshness. In either alternative of excess, whether the past be lost, or be dominant, the present is enfeebled. This is only an application of Aristotle's doctrine of the 'golden mean.'

The world is thus faced by the paradox that, at least in its higher actualities, it craves for novelty and yet is haunted by terror at the loss of the past, with its familiarities and its loved ones. It seeks escape from time in its character of 'perpetually perishing.' Part of the joy of the new years is the hope of the old round of seasons, with their stable facts—of friendship, and love, and old association. Yet conjointly with this terror, the present as mere unrelieved preservation of the past assumes the character of a horror of the past, rejection of it, revolt:

> To die, be given, or attain,
> Fierce work it were to do again.

Each new epoch enters upon its career by waging unrelenting war upon the aesthetic gods of its immediate predecessor. Yet the culminating fact of conscious, rational life refuses to conceive itself as a transient enjoyment, transiently useful. In the order of the physical world its rôle is defined by its introduction of novelty. But, just as physical feelings are haunted by the vague insistence of causality, so the higher intellectual feelings are haunted by the vague insistence of another order, where there is no unrest, no travel, no shipwreck: 'There shall be no more sea.'

This is the problem which gradually shapes itself as religion reaches its higher phases in civilized communities. The most general formulation of the religious problem is the question whether the process of the temporal world passes into the formation of other actualities, bound together in an order in which novelty does not mean loss.[1]

The ultimate evil in the temporal world is deeper than any specific evil. It lies in the fact that the past fades, that time is a 'perpetual perishing.' Objectification involves elimination. The present fact has not the past fact with it in any full immediacy. The process of time veils the past below distinctive feeling. There is a unison of becoming among things in the present. Why should there not be novelty without loss of this direct unison of immediacy among things? In the temporal world, it is the empirical fact that process entails loss: the past is present under an abstraction. But there is no reason, of any ultimate metaphysical generality, why this should be the whole story. The nature of evil is that the characters of things are mutually obstructive. Thus the depths of life require a process of selection. But the selection is elimination as the first step towards another temporal order seeking to minimize obstructive modes. Selection is at once the measure of evil, and the process of its evasion. It means discarding the element of obstructiveness in fact. No element in fact is ineffectual: thus the struggle with evil is a process of building up a mode of utilization by the provision of intermediate elements introducing a complex structure of harmony. The triviality in some initial reconstruction of order expresses the fact that actualities are being produced, which, trivial in their own proper character of immediate 'ends,' are proper 'means' for the emergence of a world at once lucid, and intrinsically of immediate worth.

The evil of the world is that those elements which are translucent so far as transmission is concerned, in themselves are of slight weight; and that those elements with individual weight, by their discord, impose upon vivid immediacy the obligation that it fade into night. 'He giveth his beloved—sleep.'

In our cosmological construction we are, therefore, left with the final opposites, joy and sorrow, good and evil, disjunction and

[1] In this paragraph Whitehead has posed the problem to be resolved in the next section. Note that in this last sentence the four polar terms are all present; order, novelty, process (or flux), and nonloss (or permanence). Note also that this problem is presented as a problem of religion and that it is characterized in the paragraphs that follow as the problem of evil.

conjunction—that is to say, the many in one—flux and permanence, greatness and triviality, freedom and necessity,[2] God and
the World. In this list, the pairs of opposites are in experience with
a certain ultimate directness of intuition, except in the case of the
last pair. God and the World introduce the note of interpretation.
They embody the interpretation of the cosmological problem in
terms of a fundamental metaphysical doctrine as to the quality of
creative origination, namely, conceptual appetition and physical
realization. This topic constitutes the last chapter of Cosmology.

II

God and the World

*This section covers four main topics: first, it presents a brief
survey and evaluation of traditional concepts of God; second, it re-
views the notion of the primordial nature of God; third, it introduces the notion of the consequent nature of God; and fourth, it
utilizes this latter concept to resolve the cosmological problem.*

1. Traditional Concepts of God

So long as the temporal world is conceived as a self-sufficient
completion of the creative act, explicable by its derivation from
an ultimate principle which is at once eminently real and the unmoved mover, from this conclusion there is no escape: the best
that we can say of the turmoil is, 'For so he giveth his beloved—
sleep.' This is the message of religions of the Buddhistic type, and
in some sense it is true. In this final discussion we have to ask,
whether metaphysical principles impose the belief that it is the
whole truth. The complexity of the world must be reflected in the
answer. It is childish to enter upon thought with the simple-
minded question, What is the world made of? The task of reason
is to fathom the deeper depths of the many-sidedness of things. We
must not expect simple answers to far-reaching questions. However
far our gaze penetrates, there are always heights beyond which
block our vision.

The notion of God as the 'unmoved mover' is derived from

[2] This pair would seem to be an alternate way of expressing novelty and order.

Aristotle, at least so far as Western thought is concerned. The notion of God as 'eminently real' is a favourite doctrine of Christian theology. The combination of the two into the doctrine of an aboriginal, eminently real, transcendent creator, at whose fiat the world came into being, and whose imposed will it obeys, is the fallacy which has infused tragedy into the histories of Christianity and of Mahometanism.

When the Western world accepted Christianity, Caesar conquered; and the received text of Western theology was edited by his lawyers. The code of Justinian and the theology of Justinian are two volumes expressing one movement of the human spirit. The brief Galilean vision of humility flickered throughout the ages, uncertainly. In the official formulation of the religion it has assumed the trivial form of the mere attribution to the Jews that they cherished a misconception about their Messiah. But the deeper idolatry, of the fashioning of God in the image of the Egyptian, Persian, and Roman imperial rulers, was retained. The Church gave unto God the attributes which belonged exclusively to Caesar.

In the great formative period of theistic philosophy, which ended with the rise of Mahometanism, after a continuance coeval with civilization, three strains of thought emerge which, amid many variations in detail, respectively fashion God in the image of an imperial ruler, God in the image of a personification of moral energy, God in the image of an ultimate philosophical principle. Hume's *Dialogues* criticize unanswerably these modes of explaining the system of the world.

The three schools of thought can be associated respectively with the divine Caesars, the Hebrew prophets, and Aristotle. But Aristotle was antedated by Indian, and Buddhistic, thought; the Hebrew prophets can be paralleled in traces of earlier thought; Mahometanism and the divine Caesars merely represent the most natural, obvious, theistically idolatrous symbolism, at all epochs and places.

The history of theistic philosophy exhibits various stages of combination of these three diverse ways of entertaining the problem. There is, however, in the Galilean origin of Christianity yet another suggestion which does not fit very well with any of the three main strands of thought. It does not emphasize the ruling

Caesar, or the ruthless moralist, or the unmoved mover. It dwells upon the tender elements in the world, which slowly and in quietness operate by love; and it finds purpose in the present immediacy of a kingdom not of this world. Love neither rules, nor is it unmoved; also it is a little oblivious as to morals. It does not look to the future; for it finds its own reward in the immediate present.

Apart from any reference to existing religions as they are, or as they ought to be, we must investigate dispassionately what the metaphysical principles, here developed, require on these points, as to the nature of God. There is nothing here in the nature of proof. There is merely the confrontation of the theoretic system with a certain rendering of the facts. But the unsystematized report upon the facts is itself highly controversial, and the system is confessedly inadequate. The deductions from it in this particular sphere of thought cannot be looked upon as more than suggestions as to how the problem is transformed in the light of that system. What follows is merely an attempt to add another speaker to that masterpiece, Hume's *Dialogues Concerning Natural Religion*. Any cogency of argument entirely depends upon elucidation of somewhat exceptional elements in our conscious experience—those elements which may roughly be classed together as religious and moral intuitions.

2. *The Primordial Nature of God*

In the first place, God is not to be treated as an exception to all metaphysical principles, invoked to save their collapse. He is their chief exemplification. Viewed as primordial, he is the unlimited conceptual realization of the absolute wealth of potentiality. In this aspect, he is not *before* all creation, but *with* all creation. But, as primordial, so far is he from 'eminent reality,' that in this abstraction he is 'deficiently actual'—and this in two ways. His feelings are only conceptual and so lack the fulness of actuality. Secondly, conceptual feelings, apart from complex integration with physical feelings, are devoid of consciousness in their subjective forms.

Thus, when we make a distinction of reason, and consider God in the abstraction of a primordial actuality, we must ascribe to

him neither fulness of feeling, nor consciousness.[3] He is the unconditioned actuality of conceptual feeling at the base of things;
so that, by reason of this primordial actuality, there is an order
in the relevance of eternal objects to the process of creation. His
unity of conceptual operations is a free creative act, untrammelled
by reference to any particular course of things. It is deflected
neither by love, nor by hatred, for what in fact comes to pass. The
particularities of the actual world presuppose *it;* while *it* merely
presupposes the *general* metaphysical character of creative advance,
of which it is the primordial exemplification. The primordial nature of God is the acquirement by creativity of a primordial
character.

His conceptual actuality at once exemplifies and establishes the
categoreal conditions. The conceptual feelings, which compose his
primordial nature, exemplify in their subjective forms their mutual
sensitivity and their subjective unity of subjective aim. These subjective forms are valuations determining the relative relevance of
eternal objects for each occasion of actuality.

He is the lure for feeling, the eternal urge of desire. His particular relevance to each creative act as it arises from its own conditioned standpoint in the world, constitutes him the initial 'object
of desire' establishing the initial phase of each subjective aim.

3.   The Consequent Nature of God

There is another side to the nature of God which cannot be
omitted. Throughout this exposition of the philosophy of organism
we have been considering the primary action of God on the world.
From this point of view, he is the principle of concretion—the
principle whereby there is initiated a definite outcome from a
situation otherwise riddled with ambiguity. Thus, so far, the
primordial side of the nature of God has alone been relevant.

But God, as well as being primordial, is also consequent. He is

---

[3] The point is, of course, that we *do* ascribe to God "eminent reality,"
"fulness of feeling," "consciousness," and concern for "what in fact comes to
pass" (see the discussion that follows). In his total being, primordial, consequent,
and superjective, he is all these. But considered by an abstraction of reason as
primordial, and primordial only, God is none of these. These are aspects of God
dependent upon his consequent and superjective natures, as will emerge in what
follows.

the beginning and the end. He is not the beginning in the sense of being in the past of all members. He is the presupposed actuality of conceptual operation, in unison of becoming with every other creative act. Thus by reason of the relativity of all things, there is a reaction of the world on God. The completion of God's nature into a fulness of physical feeling is derived from the objectification of the world in God. He shares with every new creation its actual world; and the concrescent creature is objectified in God as a novel element in God's objectification of that actual world. This prehension into God of each creature is directed with the subjective aim, and clothed with the subjective form, wholly derivative from his all-inclusive primordial valuation. God's conceptual nature is unchanged, by reason of its final completeness. But his derivative nature is consequent upon the creative advance of the world.

Thus, analogously to all actual entities, the nature of God is dipolar. He has a primordial nature and a consequent nature. The consequent nature of God is conscious; and it is the realization of the actual world in the unity of his nature, and through the transformation of his wisdom. The primordial nature is conceptual, the consequent nature is the weaving of God's physical feelings upon his primordial concepts.

One side of God's nature is constituted by his conceptual experience. This experience is the primordial fact in the world, limited by no actuality which it presupposes. It is therefore infinite, devoid of all negative prehensions. This side of his nature is free, complete, primordial, eternal, actually deficient, and unconscious. The other side originates with physical experience derived from the temporal world, and then acquires integration with the primordial side. It is determined, incomplete, consequent, 'everlasting,' fully actual, and conscious. His necessary goodness expresses the determination of his consequent nature.

Conceptual experience can be infinite, but it belongs to the nature of physical experience that it is finite. An actual entity in the temporal world is to be conceived as originated by physical experience with its process of completion motivated by consequent, conceptual experience initially derived from God. God is to be conceived as originated by conceptual experience with his process of completion motivated by consequent, physical experience, initially derived from the temporal world.

The perfection of God's subjective aim, derived from the completeness of his primordial nature, issues into the character of his consequent nature. In it there is no loss, no obstruction. The world is felt in a unison of immediacy. The property of combining creative advance with the retention of mutual immediacy is what is meant by the term 'everlasting.'[4]

The wisdom of God's subjective aim prehends every actuality for what it can be in such a perfected system—its sufferings, its sorrows, its failures, its triumphs, its immediacies of joy—woven by rightness of feeling into the harmony of the universal feeling, which is always immediate, always many, always one, always with novel advance, moving onward and never perishing. The revolts of destructive evil, purely self-regarding, are dismissed into their triviality of merely individual facts; and yet the good they did achieve in individual joy, in individual sorrow, in the introduction of needed contrast, is yet saved by its relation to the completed whole. The image—and it is but an image—the image under which this operative growth of God's nature is best conceived, is that of a tender care that nothing be lost.

The consequent nature of God is his judgment on the world. He saves the world as it passes into the immediacy of his own life. It is the judgment of a tenderness which loses nothing that can be saved. It is also the judgment of a wisdom which uses what in the temporal world is mere wreckage.

Another image which is also required to understand his consequent nature, is that of his infinite patience. The universe includes a threefold creative act composed of (i) the one infinite conceptual realization, (ii) the multiple solidarity of free physical realizations in the temporal world, (iii) the ultimate unity of the multiplicity of actual fact with the primordial conceptual fact.[5] If we conceive the first term and the last term in their unity over against the intermediate multiple freedom of physical realizations in the tem-

---

[4] This explication of the meaning attached to the word "everlasting" by Whitehead should be noted carefully; the concept figures prominently in what follows.

[5] It should be clear that (i) is the primordial nature of God, (ii) the totality of actual occasions in the temporal world, and (iii) the consequent nature of God.

poral world, we conceive of the patience of God, tenderly saving the turmoil of the intermediate world by the completion of his own nature. The sheer force of things lies in the intermediate physical process: this is the energy of physical production. God's rôle is not the combat of productive force with productive force, of destructive force with destructive force; it lies in the patient operation of the overpowering rationality of his conceptual harmonization. He does not create the world, he saves it; or, more accurately, he is the poet of the world, with tender patience leading it by his vision of truth, beauty, and goodness.

4. Resolution of the Cosmological Problem

The vicious separation of the flux from the permanence leads to the concept of an entirely static God, with eminent reality, in relation to an entirely fluent world, with deficient reality. But if the opposites, static and fluent, have once been so explained as separately to characterize diverse actualities, the interplay between the thing which is static and the things which are fluent involves contradiction at every step in its explanation. Such philosophies must include the notion of 'illusion' as a fundamental principle— the notion of '*mere* appearance.' This is the final platonic problem.

Undoubtedly, the intuitions of Greek, Hebrew, and Christian thought have alike embodied the notions of a static God condescending to the world, and of a world *either* thoroughly fluent, *or* accidentally static, but finally fluent—'heaven and earth shall pass away.' In some schools of thought, the fluency of the world is mitigated by the assumption that selected components in the world are exempt from this final fluency, and achieve a static survival. Such components are not separated by any decisive line from analogous components for which the assumption is not made. Further, the survival is construed in terms of a final pair of opposites, happiness for some, torture for others.

Such systems have the common character of starting with a fundamental intuition which we do mean to express, and of entangling themselves in verbal expressions, which carry consequences at variance with the initial intuition of permanence in fluency and of fluency in permanence.

But civilized intuition has always, although obscurely, grasped the problem as double and not as single. There is not the mere problem of fluency *and* permanence. There is the double problem: actuality with permanence, requiring fluency as its completion; and actuality with fluency, requiring permanence as its completion. The first half of the problem concerns the completion of God's primordial nature by the derivation of his consequent nature from the temporal world. The second half of the problem concerns the completion of each fluent actual occasion by its function of objective immortality, devoid of 'perpetual perishing,' that is to say, 'everlasting.'[6]

This double problem cannot be separated into two distinct problems. Either side can only be explained in terms of the other. The consequent nature of God is the fluent world become 'everlasting' by its objective immortality in God. Also the objective immortality of actual occasions requires the primordial permanence of God, whereby the creative advance ever re-establishes itself endowed with initial subjective aim derived from the relevance of God to the evolving world.

But objective immortality within the temporal world does not solve the problem set by the penetration of the finer religious intuition. 'Everlastingness' has been lost; and 'everlastingness' is the content of that vision upon which the finer religions are built —the 'many' absorbed everlastingly in the final unity.[7] The problems of the fluency of God and of the everlastingness of passing experience are solved by the same factor in the universe. This factor is the temporal world perfected by its reception and its reformation, as a fulfilment of the primordial appetition which is the basis of all order. In this way God is completed by the individual, fluent satisfactions of finite fact, and the temporal occasions are completed by their everlasting union with their transformed

---

[6] This paragraph contains the broad outline of Whitehead's solution to the cosmological problem—God's primordial nature completed by the world, and the world completed by God's consequent nature. Hence the title of the chapter, "God and the World." The remainder of the chapter is concerned with filling in the details of this statement and drawing out its significance.

[7] See fn. 4.

selves, purged into conformation with the eternal order which is the final absolute 'wisdom.' The final summary can only be expressed in terms of a group of antitheses, whose apparent self-contradictions depend on neglect of the diverse categories of existence. In each antithesis there is a shift of meaning which converts the opposition into a contrast.

It is as true to say that God is permanent and the World fluent, as that the World is permanent and God is fluent.

It is as true to say that God is one and the World many, as that the World is one and God many.

It is as true to say that, in comparison with the World, God is actual eminently, as that, in comparison with God, the World is actual eminently.

It is as true to say that the World is immanent in God, as that God is immanent in the World.

It is as true to say that God transcends the World, as that the World transcends God.

It is as true to say that God creates the World, as that the World creates God.[8]

God and the World are the contrasted opposites in terms of which Creativity achieves its supreme task of transforming disjoined multiplicity, with its diversities in opposition, into concrescent unity, with its diversities in contrast. In each actuality there are two concrescent poles of realization—'enjoyment' and 'appetition,' that is, the 'physical' and the 'conceptual.' For God the conceptual is prior to the physical, for the World the physical poles are prior to the conceptual poles.

A physical pole is in its own nature exclusive, bounded by contradiction: a conceptual pole is in its own nature all-embracing,

---

[8] These antitheses are not really as perplexing as they might at first appear. The first resolves into the statement that as primordial God is permanent while as consequent he is fluent, whereas in its immediate, temporal actuality the World is fluent while as "saved" in God's consequent nature it is permanent. In all the antitheses God is referred to once as primordial and once as consequent, and in the first, second, and fourth "the World" means in one instance its immediate temporal actuality and in the other its everlasting status in God's consequent nature—in the rest "the World" means its immediate temporal actuality in both instances. The paragraphs that follow are in large part an explication of these antitheses.

unbounded by contradiction. The former derives its share of infinity from the infinity of appetition; the latter derives its share of limitation from the exclusiveness of enjoyment. Thus, by reason of his priority of appetition, there can be but one primordial nature for God; and, by reason of their priority of enjoyment, there must be one history of many actualities in the physical world.

God and the World stand over against each other, expressing the final metaphysical truth that appetitive vision and physical enjoyment have equal claim to priority in creation. But no two actualities can be torn apart: each is all in all. Thus each temporal occasion embodies God, and is embodied in God. In God's nature, permanence is primordial and flux is derivative from the World: in the World's nature, flux is primordial and permanence is derivative from God.[9] Also the World's nature is a primordial datum for God; and God's nature is a primordial datum for the World. Creation achieves the reconciliation of permanence and flux when it has reached its final term which is everlastingness—the Apotheosis of the World.

Opposed elements stand to each other in mutual requirement. In their unity, they inhibit or contrast. God and the World stand to each other in this opposed requirement. God is the infinite ground of all mentality, the unity of vision seeking physical multiplicity. The World is the multiplicity of finites, actualities seeking a perfected unity. Neither God, nor the World, reaches static completion. Both are in the grip of the ultimate metaphysical ground, the creative advance into novelty. Either of them, God and the World, is the instrument of novelty for the other.

In every respect God and the World move conversely to each other in respect to their process. God is primordially one, namely, he is the primordial unity of relevance of the many potential forms: in the process he acquires a consequent multiplicity, which the primordial character absorbs into its own unity. The World is primordially many, namely, the many actual occasions with their physical finitude; in the process it acquires a consequent unity, which is a novel occasion and is absorbed into the multiplicity of the primordial character. Thus God is to be conceived as one and

9 Note how this sentence explicates the first antithesis.

as many in the converse sense in which the World is to be conceived as many and as one.[10] The theme of Cosmology, which is the basis of all religions, is the story of the dynamic effort of the World passing into everlasting unity, and of the static majesty of God's vision, accomplishing its purpose of completion by absorption of the World's multiplicity of effort.

* * *

'Coherence' means that the fundamental ideas, in terms of which the scheme is developed, presuppose each other so that in isolation they are meaningless. This requirement does not mean that they are definable in terms of each other; it means that what is indefinable in one such notion cannot be abstracted from its relevance to the other notions. It is the ideal of speculative philosophy that its fundamental notions shall not seem capable of abstraction from each other. In other words, it is presupposed that no entity can be conceived in complete abstraction from the system of the universe, and that it is the business of speculative philosophy to exhibit this truth. This character is its coherence.[11]

Every categoreal type of existence in the world presupposes the other types in terms of which it is explained. Thus the many eternal objects conceived in their bare isolated multiplicity lack any existent character. They require the transition to the conception of them as efficaciously existent by reason of God's conceptual realization of them.

But God's conceptual realization is nonsense if thought of under the guise of a barren, eternal hypothesis. It is God's conceptual realization performing an efficacious rôle in multiple unifications of the universe, which are free creations of actualities arising out of decided situations. Again this discordant multiplicity of actual things, requiring each other and neglecting each other, utilizing and discarding, perishing and yet claiming life as obstinate matter of fact, requires an enlargement of the understanding to the com-

[10] Note how these sentences explicate the second antithesis.

[11] Whitehead now proceeds to illustrate how his system meets this requirement of coherence. The fundamental terms introduced in illustration are eternal objects, the primordial nature of God, the temporal world of actual occasions, and the consequent nature of God.

prehension of another phase in the nature of things. The consequent nature of God is the fulfilment of his experience by his reception of the multiple freedom of actuality into the harmony of his own actualization. It is God as really actual, completing the deficiency of his mere conceptual actuality.

In this later phase, the many actualities are one actuality, and the one actuality is many actualities. Each actuality has its present life and its immediate passage into novelty; but its passage is not its death. This final phase of passage in God's nature is ever enlarging itself. In it the complete adjustment of the immediacy of joy and suffering reaches the final end of creation. This end is existence in the perfect unity of adjustment as means, and in the perfect multiplicity of the attainment of individual types of self-existence. The function of being a means is not disjoined from the function of being an end. The sense of worth beyond itself is immediately enjoyed as an overpowering element in the individual self-attainment. It is in this way that the immediacy of sorrow and pain is transformed into an element of triumph. This is the notion of redemption through suffering, which haunts the world. It is the generalization of its very minor exemplification as the aesthetic value of discords in art.

Thus the universe is to be conceived as attaining the active self-expression of its own variety of opposites—of its own freedom and its own necessity, of its own multiplicity and its own unity, of its own imperfection and its own perfection. All the 'opposites' are elements in the nature of things, and are incorrigibly there. The concept of 'God' is the way in which we understand this incredible fact—that what cannot be, yet is.

* * *

Thus the consequent nature of God is composed of a multiplicity of elements with individual self-realization. It is just as much a multiplicity as it is a unity; it is just as much one immediate fact as it is an unresting advance beyond itself. Thus the actuality of God must also be understood as a multiplicity of actual components in process of creation. This is God in his function of the kingdom of heaven.

Each actuality in the temporal world has its reception into God's nature. The corresponding element in God's nature is not temporal actuality, but is the transmutation of that temporal actuality into a living, ever-present fact. An enduring personality in the temporal world is a route of occasions in which the successors with some peculiar completeness sum up their predecessors. The correlate fact in God's nature is an even more complete unity of life in a chain of elements for which succession does not mean loss of immediate unison. This element in God's nature inherits from the temporal counterpart according to the same principle as in the temporal world the future inherits from the past. Thus in the sense in which the present occasion is the person *now,* and yet with his own past, so the counterpart in God is that person in God.

But the principle of universal relativity is not to be stopped at the consequent nature of God. This nature itself passes into the temporal world according to its gradation of relevance to the various concrescent occasions. There are thus four creative phases in which the universe accomplishes its actuality. There is first the phase of conceptual origination, deficient in actuality, but infinite in its adjustment of valuation.[12] Secondly, there is the temporal phase of physical origination, with its multiplicity of actualities.[13] In this phase full actuality is attained; but there is deficiency in the solidarity of individuals with each other. This phase derives its determinate conditions from the first phase. Thirdly, there is the phase of perfected actuality, in which the many are one everlastingly, without the qualification of any loss either of individual identity or of completeness of unity.[14] In everlastingness, immediacy is reconciled with objective immortality. This phase derives the conditions of its being from the two antecedent phases. In the fourth phase, the creative action completes itself. For the perfected actuality passes back into the temporal world, and qualifies this world so that each temporal actuality includes it as an immediate fact of relevant experience. For the kingdom of heaven

12 This is, of course, the primordial nature of God.
13 This is the temporal world of actual occasions.
14 This is the consequent nature of God.

is with us today.[15] The action of the fourth phase is the love of God for the world. It is the particular providence for particular occasions. What is done in the world is transformed into a reality in heaven, and the reality in heaven passes back into the world. By reason of this reciprocal relation, the love in the world passes into the love in heaven, and floods back again into the world. In this sense, God is the great companion—the fellow-sufferer who understands.

We find here the final application of the doctrine of objective immortality. Throughout the perishing occasions in the life of each temporal Creature, the inward source of distaste or of refreshment (the judge arising out of the very nature of things, redeemer or goddess of mischief)[16] is the transformation of Itself, everlasting in the Being of God. In this way, the insistent craving is justified— the insistent craving that zest for existence be refreshed by the ever-present, unfading importance of our immediate actions, which perish and yet live for evermore.

---

[15] This is the "superjective" nature of God. Whitehead discussed briefly the threefold character of God (primordial, consequent, superjective) in Section II, Sub-section 2 of Chapter Two.

The superjective character of God is essential, it has been argued, for the whole theory of objectification in Whitehead's system. How, it might be asked, can any past, "perished" occasion be a datum for a concrescing subject? Past occasions no longer exist in their subjective immediacy; the only reasons are *actual* occasions; but past occasions are no longer actual, being instead, objectively immortal, drained of actuality. The ground of the givenness of the past is, so the argument goes, God's superjective character. William Christian states the problem and resolves it in this fashion in his excellent study, *An Interpretation of Whitehead's Metaphysics* (New Haven, 1959), pp. 319–330. The reader who wishes critical insight into this and many other problems raised by Whitehead's cosmology is encouraged to read Christian's book, which is the most penetrating study of Whitehead yet to appear.

[16] The parentheses have been added for ease of reading. Whitehead is here indicating the specifically religious function of God, over against the metaphysical functions that have primarily concerned him. The "temporal Creatures" referred to would seem not to be individual actual occasions, but human beings constituted by routes of "perishing occasions." Therefore the "Itself" would also refer to each human being. The final sentence corroborates this interpretation.

Appendix

IN DEFENSE OF
SPECULATIVE
PHILOSOPHY

Metaphysics, in the sense of speculative philosophy, has been roundly attacked by many contemporary philosophers. The question whether speculative philosophy is a method productive of important knowledge is being vigorously discussed. In Process and Reality *Whitehead offers a most cogent defense of metaphysics. The essence of that defense follows. Not only are these arguments important in their own right as bearing upon this important question, they also serve to put Whitehead's system, which has now been laid out in some detail, into proper perspective by indicating clearly what Whitehead intended to accomplish by creating it and how he conceives of its limitations. The final two paragraphs of this appendix capture in a unique way the spirit of Whitehead the philosopher and Whitehead the man.*

This course of lectures is designed as an essay in Speculative Philosophy. Its first task must be to define 'speculative philosophy,' and to defend it as a method productive of important knowledge.

Speculative Philosophy is the endeavour to frame a coherent, logical, necessary system of general ideas in terms of which every element of our experience can be interpreted. By this notion of 'interpretation' I mean that everything of which we are conscious, as enjoyed, perceived, willed, or thought, shall have the character of a particular instance of the general scheme. Thus the philosophical scheme should be coherent, logical, and, in respect to its interpretation, applicable and adequate. Here 'applicable' means that some items of experience are thus interpretable, and 'adequate' means that there are no items incapable of such interpretation.

Philosophers can never hope finally to formulate these metaphysical first principles. Weakness of insight and deficiencies of language stand in the way inexorably. Words and phrases must be stretched towards a generality foreign to their ordinary usage; and however such elements of language be stabilized as technicalities, they remain metaphors mutely appealing for an imaginative leap.

There is no first principle which is in itself unknowable, not to be captured by a flash of insight. But, putting aside the difficulties of language, deficiency in imaginative penetration forbids progress in any form other than that of an asymptotic approach to a scheme of principles, only definable in terms of the ideal which they should satisfy.

The difficulty has its seat in the empirical side of philosophy. Our datum is the actual world, including ourselves; and this actual world spreads itself for observation in the guise of the topic of our immediate experience. The elucidation of immediate experience is the sole justification for any thought; and the starting point for thought is the analytic observation of components of this experience. But we are not conscious of any clear-cut complete analysis of immediate experience, in terms of the various details which comprise its definiteness. We habitually observe by the method of difference. Sometimes we see an elephant, and sometimes we do not. The result is that an elephant, when present, is noticed. The metaphysical first principles [however] can never fail of exemplification. We can never catch the actual world taking a holiday from their sway. Thus, for the discovery of metaphysics, the method of pinning down thought to the strict systematization of detailed discrimination, already effected by antecedent observation, breaks down. This collapse of the method of rigid empiricism is not confined to metaphysics. It occurs whenever we seek the larger generalities.

In natural science this rigid method is the Baconian method of induction, a method which, if consistently pursued, would have left science where it found it. What Bacon omitted was the play of a free imagination, controlled by the requirements of coherence and logic. The true method of discovery is like the flight of an aeroplane. It starts from the ground of particular observation; it makes a flight in the thin air of imaginative generalization; and it again lands for renewed observation rendered acute by rational interpretation. The reason for the success of this method of imaginative rationalization is that, when the method of difference fails, factors which are constantly present may yet be observed under the influence of imaginative thought. Such thought supplies the differences which the direct observation lacks. It can even play with inconsistency; and can thus throw light on the consistent, and persistent, elements in experience by comparison with what in imagination is inconsistent with them. The negative judgment is the peak of mentality.

But the conditions for the success of imaginative construction must be rigidly adhered to. In the first place, this construction must have its origin in the generalization of particular factors discerned in particular topics of human interest; for example, in physics, or in physiology, or in psychology, or in aesthetics, or in ethical beliefs, or in sociology, or in languages conceived as storehouses of human experience. In this way the prime requisite, that anyhow there shall be some important application, is secured. The success of the imaginative experiment is always to be tested by the applicability of its results beyond the restricted locus from which it originated. In default of such extended application, a generalization started from physics, for example, remains merely an alternative expression of notions applicable to physics. The partially successful philosophic generalization will, if derived from physics, find applications in fields of experience beyond physics. It will enlighten observation in those remote fields, so that general principles can be discerned as in process of illustration, which in the absence of the imaginative generalization are obscured by their persistent exemplification.

Thus the first requisite is to proceed by the method of generalization so that certainly there is some application; and the test of some success is application beyond the immediate origin. In other words, some synoptic vision has been gained.

In this description of philosophic method, the term 'philosophic generalization' has meant 'the utilization of specific notions, applying to a restricted group of facts, for the divination of the generic notions which apply to all facts.'

In its use of this method natural science has shown a curious mixture of rationalism and irrationalism. Its prevalent tone of thought has been ardently rationalistic within its own borders, and dogmatically irrational beyond those borders. In practice such an attitude tends to become a dogmatic denial that there are any factors in the world not fully expressible in terms of its own primary notions devoid of further generalization. Such a denial is the self-denial of thought.

The second condition for the success of imaginative construction is unflinching pursuit of the two rationalistic ideals, coherence and logical perfection.

Logical perfection does not here require any detailed explanation. An example of its importance is afforded by the rôle of mathematics in the restricted field of natural science. The history of mathematics exhibits the generalization of special notions observed in particular instances.

The requirement of coherence is the great preservative of rationalistic sanity. But the validity of its criticism is not always admitted. If we consider philosophical controversies, we shall find that disputants tend to require coherence from their adversaries, and to grant dispensations to themselves. It has been remarked that a system of philosophy is never refuted; it is only abandoned. The reason is that logical contradictions, except as temporary slips of the mind—plentiful, though temporary—are the most gratuitous of errors; and usually they are trivial. Thus, after criticism, systems do not exhibit mere illogicalities. They suffer from inadequacy and incoherence. Failure to include some obvious elements of experience in the scope of the system is met by boldly denying the facts. Also while a philosophical system retains any charm of novelty, it enjoys a plenary indulgence for its failures in coherence. But after a system has acquired orthodoxy, and is taught with authority, it receives a sharper criticism. Its denials and its incoherences are found intolerable, and a reaction sets in.

Incoherence is the arbitrary disconnection of first principles. In modern philosophy Descartes' two kinds of substance, corporeal and mental, illustrate incoherence. There is, in Descartes' philosophy, no reason why there should not be a one-substance world, only corporeal, or a one-substance world, only mental. According to Descartes, a substantial individual 'requires nothing but itself in order to exist.' Thus this system makes a virtue of its incoherence. But on the other hand, the facts seem connected, while Descartes' system does not; for example, in the treatment of the body-mind problem. The Cartesian system obviously says something that is true. But its notions are too abstract to penetrate into the nature of things.

The attraction of Spinoza's philosophy lies in its modification of Descartes' position into greater coherence. He starts with one substance, *causa sui*, and considers its essential attributes and its individualized modes, i.e. the *'affectiones substantiae.'* The gap in the system is the arbitrary introduction of the 'modes.' And yet, a multiplicity of modes is a fixed requisite, if the scheme is to retain any direct relevance to the many occasions in the experienced world.

In its turn every philosophy will suffer a deposition. But the bundle of philosophic systems expresses a variety of general truths about the universe, awaiting co-ordination and assignment of their various spheres of validity. Such progress in co-ordination is provided by the advance of philosophy; and in this sense philosophy has advanced from Plato onwards. According to this account of the achievement of rationalism, the chief error in philosophy is overstatement. The aim at generalization is sound, but the estimate of success is exaggerated. There are two main forms of such overstatement. One form is what I have termed elsewhere (*Science and the Modern World*, Ch. III) the 'fallacy of misplaced concreteness.' This fallacy consists in neglecting the degree of abstraction involved when an actual entity is considered merely so far as it exemplifies certain categories of thought. There are aspects of actualities which are simply ignored so long as we restrict thought to these categories. Thus the success of a philosophy is to be measured by its comparative avoidance of this fallacy, when thought is restricted within its categories.

The other form of overstatement consists in a false estimate of logical procedure in respect to certainty, and in respect to premises. Philosophy has been haunted by the unfortunate notion that its method is dogmatically to indicate premises which are severally clear, distinct, and certain; and to erect upon those premises a deductive system of thought.

But the accurate expression of the final generalities is the goal of discussion and not its origin. Philosophy has been misled by the example of mathematics; and even in mathematics the statement of the ultimate logical principles is beset with difficulties, as yet insuperable. The verification of a rationalistic scheme is to be sought in its general success, and not in the peculiar certainty, or initial clarity, of its first principles.

If we may trust the Pythagorean tradition, the rise of European philosophy was largely promoted by the development of mathematics into a science of abstract generality. But in its subsequent development the method of philosophy has also been vitiated by the example of mathematics. The primary method of mathematics is deduction; the primary method of philosophy is descriptive generalization. Under the influence of mathematics, deduction has been foisted onto philosophy as its standard method, instead of taking its true place as an essential auxiliary mode of verification whereby to test the scope of generalities. This misapprehension of philosophic method has veiled the very considerable success of philosophy in providing generic notions which add lucidity to our apprehension of the facts of experience. The depositions of Plato, Aristotle, Thomas Aquinas, Descartes, Spinoza, Leibniz, Locke, Berkeley, Hume, Kant, Hegel, merely mean that ideas which these men introduced into the philosophic tradition must be construed with limitations, adaptations, and inversions, either unknown to them, or even explicitly repudiated by them. A new idea introduces a new alternative; and we are not less indebted to a thinker when we adopt the alternative which he discarded. Philosophy never reverts to its old position after the shock of a great philosopher.

The study of philosophy is a voyage towards the larger generalities. For this reason in the infancy of science, when the main stress lay in the discovery of the most general ideas usefully applicable to the subject-matter in question, philosophy was not sharply distinguished from science. To this day, a new science with any substantial novelty in its notions is considered to be in some way peculiarly philosophical. In their later stages, apart from occasional disturbances, most sciences accept without question the general notions in terms of which they develop. The main stress is laid on the adjustment and the direct verification of more special statements. In such periods scientists repudiate philosophy; Newton, justly satisfied with his physical principles, disclaimed metaphysics.

The fate of Newtonian physics warns us that there is a development in scientific first principles, and that their original forms can only be saved by interpretations of meaning and limitations of their field of application—interpretations and limitations unsuspected during the first period of successful employment. One chapter in the history of culture is concerned with the growth of generalities. In such a chapter it is seen that the older generalities, like the older hills, are worn down and diminished in height, surpassed by younger rivals.

Thus one aim of philosophy is to challenge the half-truths constituting the scientific first principles. The systematization of knowledge cannot be conducted in watertight compartments. All general truths condition each other; and the limits of their application cannot be adequately defined apart from their correlation by yet wider generalities. The criticism of principles must chiefly take the form of determining the proper meanings to be assigned to the fundamental notions of the various sciences, when these notions are considered in respect to their status relatively to each other. The determination of this status requires a generality transcending any special subject-matter.

Philosophy will not regain its proper status until the gradual elaboration of categoreal schemes, definitely stated at each stage of progress, is recognized as its proper objective. There may be rival schemes, inconsistent among themselves; each with its own merits and its own failures. It will then be the purpose of research to conciliate the differences. Metaphysical categories are not dogmatic statements of the obvious; they are tentative formulations of the ultimate generalities.

If we consider any scheme of philosophic categories as one complex assertion, and apply to it the logician's alternative, true or false, the answer must be that the scheme is false. The same answer must be given to a like question respecting the existing formulated principles of any science.

The scheme is true with unformulated qualifications, exceptions, limitations, and new interpretations in terms of more general notions. We do not yet know how to recast the scheme into a logical truth. But the scheme is a matrix from which true propositions applicable to particular circumstances can be derived. We can at present only trust our trained instincts as to the discrimination of the circumstances in respect to which the scheme is valid.

The use of such a matrix is to argue from it boldly and with rigid logic. The scheme should therefore be stated with the utmost precision and definiteness, to allow of such argumentation. The conclusion of the argument should then be confronted with circumstances to which it should apply.

The primary advantage thus gained is that experience is not interrogated with the benumbing repression of common sense. The observation acquires an enhanced penetration by reason of the expectation evoked by the conclusion of the argument. The outcome from this procedure takes one of three forms: (i) the conclusion may agree with the observed facts; (ii) the conclusion may exhibit general agreement, with disagreement in detail; (iii) the conclusion may be in complete disagreement with the facts.

In the first case, the facts are known with more adequacy and the applicability of the system to the world has been elucidated. In the second case, criticisms of the observation of the facts and of the details of the scheme are both required. The history of thought shows that false interpretations of observed facts enter into the records of their observation. Thus both theory, and received notions as to fact, are in doubt. In the third case a fundamental reorganization of theory is required either by way of limiting it to some special province, or by way of entire abandonment of its main categories of thought.

That we fail to find in experience any elements intrinsically incapable of exhibition as examples of general theory, is the hope of rationalism. This hope is not a metaphysical premise. It is the faith which forms the motive for the pursuit of all sciences alike, including metaphysics.

In so far as metaphysics enables us to apprehend the rationality of things, the claim is justified. It is always open to us, having regard to the imperfections of all metaphysical systems, to lose hope at the exact point where we find ourselves. The preservation of such faith must depend on an ultimate moral intuition into the nature of intellectual action—that it should embody the adventure of hope. Such an intuition marks the point where metaphysics— and indeed every science—gains assurance from religion and passes over into religion. But in itself the faith does not embody a premise from which the theory starts; it is an ideal which is seeking satisfaction. In so far as we believe that doctrine, we are rationalists. There must, however, be limits to the claim that all the elements in the universe are explicable by 'theory.' For 'theory' itself requires that there be 'given' elements so as to form the material for theorizing.

Every science must devise its own instruments. The tool required for philosophy is language. Thus philosophy redesigns language in the same way that, in a physical science, pre-existing appliances are redesigned. It is exactly at this point that the appeal to facts is a difficult operation. This appeal is not solely to the expression of the facts in current verbal statements. The adequacy of such sentences is the main question at issue. It is true that the general agreement of mankind as to experienced facts is best expressed in language. But the language of literature breaks down precisely at the task of expressing in explicit form the larger generalities—the very generalities which metaphysics seeks to express.

The point is that every proposition refers to a universe exhibiting some general systematic metaphysical character. Apart from this background, the separate entities which go to form the proposition, and the proposition as a whole, are without determinate character. Nothing has been defined, because every definite entity requires a systematic universe to supply its requisite status. Thus every proposition proposing a fact must, in its complete analysis, propose the general character of the universe required for that fact. There are no self-sustained facts, floating in nonentity.

The technical language of philosophy represents attempts of various schools of thought to obtain explicit expression of general ideas presupposed by the facts of experience. It follows that any novelty in metaphysical doctrines exhibits some measure of disagreement with statements of the facts to be found in current philosophical literature. The extent of disagreement measures the extent of metaphysical divergence. It is, therefore, no valid criticism of one metaphysical school to point out that its doctrines do not follow from the verbal expression of the facts accepted by another school. The whole contention is that the doctrines in question supply a closer approach to fully expressed propositions.

Whatever is found in 'practice' must lie within the scope of the metaphysical description. When the description fails to include the 'practice,' the metaphysics is inadequate and requires revision. There can be no appeal to practice to supplement metaphysics, so long as we remain contented with our metaphysical doctrines. Metaphysics is nothing but the description of the generalities which apply to all the details of practice.

No metaphysical system can hope entirely to satisfy these pragmatic tests. At the best such a system will remain only an approximation to the general truths which are sought. In particular, there are no precisely stated axiomatic certainties from which to start. There is not even the language in which to frame them. The only possible procedure is to start from verbal expressions which, when taken by themselves with the current meaning of their words, are ill-defined and ambiguous. These are not premises to be immediately reasoned from apart from elucidation by further discussion; they are endeavours to state general principles which will be exemplified in the subsequent description of the facts of experience. This subsequent elaboration should elucidate the meanings to be assigned to the words and phrases employed. Such meanings are incapable of accurate apprehension apart from a correspondingly accurate apprehension of the metaphysical background which the universe provides for them. But no language can be anything but elliptical, requiring a leap of the imagination to understand its meaning in its relevance to immediate experience. The position of metaphysics in the development of culture cannot be understood without remembering that no verbal statement is the adequate expression of a proposition.

An old established metaphysical system gains a false air of adequate precision from the fact that its words and phrases have passed into current literature. Thus propositions expressed in its language are more easily correlated to our flitting intuitions into metaphysical truth. When we trust these verbal statements and argue as though they adequately analysed meaning, we are led into difficulties which take the shape of negations of what in practice is presupposed. But when they are proposed as first principles they assume an unmerited air of sober obviousness. Their defect is that the true propositions which they do express lose their fundamental character when subjected to adequate expression. For example consider the type of propositions such as 'The grass is green,' and 'The whale is big.' This subject-predicate form of statement seems so simple, leading straight to a metaphysical first principle; and yet in these examples it conceals such complex, diverse meanings.

It has been an objection to speculative philosophy that it is overambitious. Rationalism, it is admitted, is the method by which advance is made within the limits of particular sciences. It is, however, held that this limited success must not encourage attempts to frame ambitious schemes expressive of the general nature of things.

One alleged justification of this criticism is ill-success: European thought is represented as littered with metaphysical systems, abandoned and unreconciled. Such an assertion tacitly fastens upon philosophy the old dogmatic test. The same criterion would fasten ill-success upon science. We no more retain the physics of the seventeenth century than we do the Cartesian philosophy of that century. Yet within limits, both systems express important truths. Also we are beginning to understand the wider categories which define their limits of correct application. Of course, in that century, dogmatic views held sway; so that the validity both of the physical notions, and of the Cartesian notions, was misconceived. Mankind never quite knows what it is after. When we survey the history of thought, and likewise the history of practice, we find that one idea after another is tried out, its limitations defined, and its core of truth elicited. In application to the instinct for the intellectual adventures demanded by particular epochs, there is much truth in Augustine's rhetorical phrase, *Securus judicat orbis terrarum*. At the very least, men do what they can in the way of systematization, and in the event achieve something. The proper test is not that of finality, but of progress.

But the main objection, dating from the sixteenth century and receiving final expression from Francis Bacon, is the uselessness of philosophic speculation. The position taken by this objection is that we ought to describe detailed matter of fact, and elicit the laws with a generality strictly limited to the systematization of these described details. General interpretation, it is held, has no bearing upon this procedure; and thus any system of general interpretation, be it true or false, remains intrinsically barren. Unfortunately for this objection, there are no brute, self-contained matters of fact, capable of being understood apart from interpretation as an element in a system. Whenever we attempt to express the matter of immediate experience, we find that its understanding leads us beyond itself, to its contemporaries, to its past, to its future, and to the universals in terms of which its definiteness is exhibited. But such universals, by their very character of universality, embody the potentiality of other facts with variant types of definiteness. Thus the understanding of the immediate brute fact requires its metaphysical interpretation as an item in a world with some systematic relation to it. When thought comes upon the scene, it finds the interpretations as matters of practice. Philosophy does not initiate interpretations. Its search for a rationalistic scheme is the search for more adequate criticism, and for more adequate justification, of the interpretations which we perforce employ.

The conclusion of this discussion is, first, the assertion of the old doctrine that breadth of thought reacting with intensity of sensitive experience stands out as an ultimate claim of existence; secondly, the assertion that empirically the development of self-justifying thoughts has been achieved by the complex process of generalizing from particular topics, of imaginatively schematizing the generalizations, and finally by renewed comparison of the imagined scheme with the direct experience to which it should apply.

There is no justification for checking generalization at any particular stage. Each phase of generalization exhibits its own peculiar simplicities which stand out just at that stage, and at no other stage. There are simplicities connected with the motion of a bar of steel which are obscured if we refuse to abstract from the individual molecules; and there are certain simplicities concerning the behaviour of men which are obscured if we refuse to abstract from the individual peculiarities of particular specimens. In the same way, there are certain general truths, about the actual things in the common world of activity, which will be obscured when attention is confined to some particular detailed mode of considering them. These general truths, involved in the meaning of every particular notion respecting the actions of things, are the subject matter for speculative philosophy.

Philosophy destroys its usefulness when it indulges in brilliant feats of explaining away. It is then trespassing with the wrong equipment upon the field of particular sciences. Its ultimate appeal is to the general consciousness of what in practice we experience. Whatever thread of presupposition characterizes social expression throughout the various epochs of rational society, must find its place in philosophic theory. Speculative boldness must be balanced by complete humility before logic, and before fact. It is a disease of philosophy when it is neither bold nor humble, but merely a reflection of the temperamental presuppositions of exceptional personalities.

Analogously, we do not trust any recasting of scientific theory depending upon a single performance of an aberrant experiment, unrepeated. The ultimate test is always widespread, recurrent experience; and the more general the rationalistic scheme, the more important is this final appeal.

The useful function of philosophy is to promote the most general systematization of civilized thought. There is a constant reaction between specialism and common sense. It is the part of the special sciences to modify common sense. Philosophy is the welding of imagination and common sense into a restraint upon specialists, and also into an enlargement of their imaginations. By providing the generic notions philosophy should make it easier to conceive the infinite variety of specific instances which rest unrealized in the womb of nature.

In these lectures I have endeavoured to compress the material derived from years of meditation. In putting out these results, four strong impressions dominate my mind: First, that the movement of historical, and philosophical, criticism of detached questions, which on the whole has dominated the last two centuries, has done its work, and requires to be supplemented by a more sustained effort of constructive thought. Secondly, that the true method of philosophical construction is to frame a scheme of ideas, the best that one can, and unflinchingly to explore the interpretation of experience in terms of that scheme. Thirdly, that all constructive thought, on the various special topics of scientific interest, is dominated by some such scheme, unacknowledged, but no less influential in guiding the imagination. The importance of philosophy lies in its sustained effort to make such schemes explicit, and thereby capable of criticism and improvement.

There remains the final reflection, how shallow, puny, and imperfect are efforts to sound the depths in the nature of things. In philosophical discussion, the merest hint of dogmatic certainty as to finality of statement is an exhibition of folly.

Glossary

Actual entity—" 'Actual entities'—also termed 'actual occasions'—are the final real things of which the world is made up" [*PR* 27]. Like the atoms of Democritus they are microcosmic entities, aggregates of which, termed *societies* or *nexūs,* form the macrocosmic entities of our everyday experience—trees, houses, people. But whereas the atoms of Democritus are inert, imperishable, material stuff, Whitehead's actual entities are vital, transient "drops of experience, complex and interdependent" [*PR* 28]. To hold that the final real things of which the world is made up are drops of experience is not to imply that consciousness permeates inanimate nature; for consciousness can characterize only extremely sophisticated actual entities, and actual entities have the potentiality for the sophistication productive of consciousness only when they are members of extremely complex societies such as the society we call the human brain. Whitehead's insight is to see that if one is to take the doctrine of evolution seriously and hold that sentient, purposive creatures gradually emerged out of the primordial ooze, then that ooze must be understood in such a way that the emergence from it of animals and human beings is intelligible. Hence he advocates a neutral monism in which actual entities are neither bits of material stuff nor Leibnizian souls, but, rather, units of process that may be linked to other actual entities to form temporal strands of matter, or perhaps linked with other sophisticated actual entities, all of which are intricately involved with a complex society like a brain, to form a route of inheritance that we identify as the conscious soul of an enduring person. (*See* STRUCTURED SOCIETY *for a fuller account.*)

Actual entities, then, are units of process, and the title *Process and Reality* is meant to indicate that for Whitehead these microcosmic units of process are the final realities—"there is no going behind actual entities to find anything more real" [*PR* 27–28]. On the other hand, to mistakenly consider an aggregate of actual entities as a final reality is to commit the Fallacy of Misplaced Concreteness; Descartes was guilty of this fallacy when he identified mind and matter as two distinct kinds of reality.

An actual entity is "the unity to be ascribed to a particular instance of concrescence" [*PR* 323]. A concrescence is a growing together of the remnants of the perishing past into the vibrant immediacy of a novel, present unity. An actual entity endures but an instant—the instant of its becoming, its active process of self-creation out of the elements of the perishing past—and then it, too, perishes and as objectively immortal becomes dead datum for succeeding generations of actual entities. The concrescence of an actual entity begins with a passive, receptive moment when the givenness of the past is thrust upon it; it then completes its becoming through a series of creative supplemental phases that adjust, integrate, and perhaps modify the given data. In simple actual entities there is mere reiteration of the given; they are "vehicles for receiving, for storing in a napkin, and for restoring without loss or gain" [*PR* 269]. Sophisticated actual entities enjoy a complex inheritance as a result of their social involvement, and this complexity of inheritance begets originality in the supplemental phases as the means for achieving integration and unity.

Actual occasion—For all practical purposes the phrases *actual occasion* and *actual entity* are interchangeable. Whitehead notes only one difference: the word *occasion* implies a spatio-temporal location. God is the one nontemporal actual entity. Hence Whitehead observes that "the term 'actual occasion' will always exclude God from its scope" [*PR* 135]. It is true, however, that even though "the term 'actual occasion' is used synonymously with 'actual entity' " [*PR* 119], the use of *actual occasion* should alert one to the likelihood that the "character of extensiveness has some direct relevance to the discussion, either extensiveness in the form of temporal extensiveness, that is to say 'duration,' or extensiveness in the form of spatial extension, or in the more complete signification of spatio-temporal extensiveness" [*PR* 119].

Actual world—The actual world of any particular actual entity is that collection of actual entities which are objectified as given data for the initial, receptive phase of the concrescence of that particular actual entity. The phrase *actual world,* like the phrase *the present,* is relative—i.e., it shifts its meaning with every shift of standpoint; no two actual entities experience identical actual worlds.

Affirmation-negation contrast—A contrast is the experienced unity of the several different components in a complex datum. The affirmation-negation contrast is a holding together in a unity of two very special kinds of components, and this particular kind of contrast has been singled out by giving it a special name for the important reason that consciousness is the subjective form involved in feeling this kind of contrast.

In the affirmation-negation contrast the two components held together in a unity—i.e., synthesized in one datum—are (1) a feeling of a nexus of actual entities and (2) a feeling of a proposition with its logical subjects members of the nexus. Whitehead calls propositions *theories,* so this contrast can be called that between fact that is given and theory, related to that fact, which may be erroneous. In Figure 2 in this book, bracket *z* represents pictorially an affirmation-negation contrast, and circle *d* represents the feeling of that contrast, which could involve consciousness in its subjective form.

Appetition—"Appetition is at once the conceptual valuation of an immediate physical feeling combined with the urge towards realization of the datum conceptually prehended" [*PR* 47]. According to the fourth categoreal obligation there is a purely conceptual feeling derived from each physical feeling, and the subjective form of a conceptual feeling has the character of a valuation, which is either valuation upward (adversion) or valuation downward (aversion). Appetition "is immediate matter of fact including in itself a principle of unrest" [*PR* 47], and this element of valuation—adversion or aversion—is the locus of unrest. By means of reversion there may be novel conceptual feelings present in the mental pole of an actual entity, and by means of appetition such an actual entity "thereby conditions creativity so as to procure, in the future, physical realization of its mental pole" [*PR* 48].

Appetition is also used in connection with the primordial nature of God. God's primordial nature is his "unfettered conceptual valuation, 'infinite' in Spinoza's sense of that term" [*PR* 378], of the realm of eternal objects. Note that this is not simply conceptual feeling; it is conceptual valuation. "This is the ultimate, basic adjustment of the togetherness of eternal objects on which creative order depends" [*PR* 48]. "The graduated order of appetitions constituting the primordial nature of God" [*PR* 315] is prehended by every actual entity, for each of which it is a lure towards which the actual entity may aim its concrescence. God's purpose is to promote intensity of feeling; his appetitions constitute his purpose, and as felt in the world they implement his purpose. (*See also* SUBJECTIVE AIM.)

Causal efficacy—Causal efficacy is the more primitive and fundamental of the two pure modes of perception. (Presentational immediacy is the other pure mode, and both combine to form the mixed mode of perception, symbolic reference, which is our ordinary mode of awareness.) Causal efficacy, as a pure mode of perception, does not involve consciousness or life, but is present in all actual entities whatsoever, including those constitutive of inanimate material objects.

Perception in the mode of causal efficacy is the basic mode of inheritance of feeling from past data, and the feelings it transmits are vague, massive, inarticulate, and felt as the efficaciousness of the past. It is what Whitehead refers to as *crude* perception, and it arises in the first phase of concrescence as conformal feeling. Conformal feelings have a vector character; they are the agency by which other things are built into—i.e., objectified for—any given subject in process of conscrescence and are therefore also called *causal feelings.* "The 'power' of one actual entity on the other is simply how the former is objectified in the constitution of the other" [*PR* 91]. (*See also* PRESENTATIONAL IMMEDIACY *and* SYMBOLIC REFERENCE.)

Causation: efficient and final—"One task of a sound metaphysics is to exhibit final and efficient causes in their proper relation to each other" [*PR* 129]. Aristotle's interest in biology resulted in principles that "led to a wild overstressing of the notion of final causes during the Christian ages; and thence, by a reaction, to the correlative overstressing of the notion of 'efficient causes' during the modern scientific period" [*PR* 128–129]. Whitehead strives to reach the mean between these extremes.

The relationship between efficient and final causation in Whitehead's philosophy is the relationship between the initial, conformal phase of concrescence and the succeeding supplemental phases. The initial phase is the phase of efficient causation, of causal efficacy, and the supplemental phases are, if they are at all significant, the phases of novelty and purposive adjustment. In the simple occasions constitutive of inanimate material objects the supplemental phases are impotent and efficient causation reigns supreme, but in the case of more sophisticated actual entities, where the supplemental phases are significant, a description purely in terms of efficient causation abstracts from the full reality of the actual entities involved and results in distortion. Supplemental activity can range from being trivial to being of over-riding importance, as in the case of the routes of regnant actual entities in men and animals.

"The doctrine of the philosophy of organism is that, however far the sphere of efficient causation be pushed in the determination of components of a conscrescence . . . beyond the determination of these components there always remains the final reaction of the self-creative unity of the universe. This final reaction completes the self-creative act by putting the decisive stamp of creative emphasis upon the determinations of efficient cause. Each occasion exhibits its measure of creative emphasis in proportion to its measure of subjective intensity. . . . But in the temporal world for occasions of relatively slight experiential intensity, their decisions of creative emphasis are individually negligible compared to the determined components which they receive and transmit" [*PR* 75].

Change—Actual entities never change nor move; they become, and their becoming is also their moment of perishing. "An actual entity never moves: it is where it is and what it is" [*PR* 113]. But then, as Whitehead notes, it is "quite obvious that meanings have to be found for the notions of 'motion' and of 'moving bodies'" [*PR* 113]. The meanings are framed in terms of the notion of an event. An event is a nexus of actual entities, an extensive string of interrelated actual entities. "The fundamental meaning of the notion of 'change' is 'the difference between actual occasions comprised in some determinate event'" [*PR* 114]. Another, more basic, way of putting this is in terms of eternal objects: "'Change' is the description of the adventures of eternal objects in the evolving universe of actual things" [*PR* 92]. These two accounts presuppose each other; if there were no modifications in the patterns of eternal objects present in successive actual entities constitutive of an event, then there would be no change, no motion, but the mere duration resulting from reiteration.

Conceptual prehension—A conceptual prehension is a prehension whose datum is an eternal object. Even though physical prehensions are prehensions whose data are actual entities, they still involve eternal objects, but eternal objects as immanent—i.e., as determined to specific actualities, namely, those actual entities being prehended. By contrast, a conceptual prehension "is the feeling of an unqualified negation; that is to say, it is the feeling of a definite eternal object with the definite extrusion of any particular realization" [*PR* 372].

As indicated in Figure 2 of this book, the second phase of concrescence is the phase of conceptual prehensions. Each actual entity initiates its concrescence with a phase of conformal, physical feelings and by Categoreal Obligation IV, "from each physical feeling there is the derivation of a purely conceptual feeling whose datum is the eternal object exemplified in the definiteness of the actual entity, or the nexus, physically felt" [*PR* 379]. With the emergence of conceptual prehensions there is, because of their character as unqualified negations, the possibility of liberation from the tyranny of the given. The subjective form of a conceptual feeling is valuation—either valuation up (adversion) or valuation down (aversion). This conceptual valuation, which either enhances or inhibits the importance of its datum eternal object in the working out of the concrescence, is the most primitive type of creative response open to actual entities; here at this level creative purpose is already operative and the concrescing subject is acting autonomously as a determinant of its own concrescence, though it is true that the later phases of comparative prehensions offer much more latitude for creative response.

Conceptual reversion—The second phase of concrescence, the phase of conceptual prehensions (*see Figure 2 of this book*), reiterates conceptually the eternal objects that, in the first phase of concrescence, have been experienced as characterizing datum actual entities. But this second phase of concrescence is in actuality composed of two distinguishable subphases, conceptual reiteration and conceptual reversion. Reiteration is always a factor in concrescence; reversion may or may not occur.

Conceptual reversion is "secondary origination of conceptual feelings with data which are partially identical with, and partially diverse from, the eternal objects forming the data in the primary [subphase of phase two of concrescence]" [*PR* 380]. As a result of conceptual reversion "the proximate novelties are conceptually felt. This is the process by which the subsequent enrichment of subjective forms, both in qualitative pattern, and in intensity through contrast, is made possible by the positive conceptual prehension of relevant alternatives. . . . [Conceptual reversion] is the category by which novelty enters the world; so that even amid stability there is never undifferentiated endurance" [*PR* 381].

All conceptual feeings are feelings whose data are eternal objects. Some conceptual feelings merely reiterate, conceptually, the forms of definiteness felt physically in the first phase of physical feelings, but other conceptual feelings have as data eternal objects that have not been felt as characterizing fact, but have arisen by means of reversion. These eternal objects that have become relevant as a result of reversion have, however, something in common with the reiterated eternal objects; their novelty is a relevant novelty. Relevance is determined by the aim at contrast governed by the final aim at intensity of feeling that is "the expression of the ultimate creative purpose that each unification shall achieve some maximum depth of intensity of feeling . . . ." [*PR* 381]. This ultimate purpose is, in the last analysis, the purpose of God, and "a more fundamental account must ascribe the reverted conceptual feeling in a temporal subject to its conceptual feeling derived . . . from the hybrid physical feeling of the relevancies conceptually ordered in God's experience" [*PR* 382]. Reversion, therefore, is a function of each actual entity's feeling of God's primordial nature, which is the "primordial unity of relevance of the many potential forms" [*PR* 529]. (*See also* SUBJECTIVE AIM.)

Concrescence—*Concrescence* is the name given to the process that *is* any given actual entity; it is "the real internal constitution of a particular existent" [*PR* 320]. Concrescence is the *growing together* of a many into the unity of a one. (*See* CREATIVITY.) The initial phase of a concrescence is composed of the separate feelings of the disjunctively diverse entities that make up the actual world of the actual entity in question. Subsequent phases effect the *growing together,* the *concrescence,* of these many separate feelings into one unity of feeling, which is termed the *satisfaction* of that actual entity. " 'Concrescence' is the name for the process in which the universe of many things acquires an individual unity in a determinate relegation of each item of the 'many' to its subordination in the constitution of the novel 'one' " [PR 321]. With the attaining of its satisfaction an actual entity is completed and *perishes*—i.e., it becomes a datum for fresh instances of concrescence. (*See also* process.)

Conformal feelings—The first phase of concrescence is called indifferently the *conformal phase,* the *responsive phase,* the *initial phase,* the *receptive phase,* or the *primary phase,* and the prehensions constitutive of this phase are called *conformal feelings, responsive feelings,* or *pure physical feelings.* The conformal phase initiates a concrescence by providing a link between the past and the present. The past is given as objectified data, and "the responsive phase absorbs these data as material for a subjective unity of feeling" [PR 261]. These conformal feelings "are 'vectors'; for they feel what is *there* and transform it into what is *here*" [PR 133]. A conformal feeling "transforms the objective content into subjective feelings" [PR 250]. Conformal feelings are the means by which "the datum, which is mere potentiality, becomes the individualized basis for a complex unity of realization" [PR 173]; though the unity is not achieved until later phases of concrescence, the materials from which such a unity will be forged are provided by the conformal feelings.

Conformal feelings reiterate, reproduce—the comparative feelings of the later, supplemental phases are the source of novelty, not the conformal feelings. It is true, however, that even at the level of conformal feelings there is creative selectivity under the guidance of subjective aim, for some elements of the datum occasion of any conformal feeling are selected for objectification and some are eliminated from efficacy by negative prehensions.

The following sentences indicate the importance in Whitehead's scheme of conformal feelings. "This direct perception, characterized by mere subjective responsiveness and by lack of origination in the higher phases, exhibits the constitution of an actual entity under the guise of receptivity. In the language of causation, it describes the efficient causation operative in the actual world. In the language of epistemology, as framed by Locke, it describes how the ideas of particular existents are absorbed into the subjectivity of the percipient and are the datum for its experience of the external world. In the language of science, it describes how the quantitative intensity of localized energy bears in itself the vector marks of its origin, and the specialities of its specific forms; it also gives a reason for the atomic quanta to be discerned in the building up of a quantity of energy. In this way, the philosophy of organism—as it should— appeals to the facts" [*PR* 178–179].

Consciousness—"Consciousness is how we feel the affirmation-negation contrast" [*PR* 372]. An affirmation-negation contrast involves holding together in a unity as one datum a feeling of a nexus of actual entities and a feeling of a proposition with its logical subjects members of the nexus. (*See* AFFIRMATION-NEGATION CONTRAST *for a fuller account.*)

Because propositional feelings occur in the third phase of concrescence, and because the contrast between propositional feelings and fact that involves consciousness is a product of even higher integrations in the fourth phase of concrescence, it follows that consciousness "is the crown of experience, only occasionally attained, not its necessary base" [*PR* 408]. Because "consciousness presupposes experience and not experience consciousness" [*PR* 83], it is the case that "clearness in consciousness is no evidence for primitiveness in the genetic process: the opposite doctrine is more nearly true" [*PR* 264–265]. This set of relationships, central to Whitehead's system, is at the heart of his attack on Hume, for Whitehead argues that Hume's whole analysis of causation is irrelevant from the beginning because Hume started with conscious perception as his primary fact and argued that "any apprehension of causation was, somehow or other, to be elicited from this primary fact" [*PR* 263]. Of course Hume could never elicit such an apprehension, but, Whitehead would argue, this proves nothing about causation because causal relationships are to be found in the primitive prehensions that consciousness presupposes but only vaguely apprehends. Whitehead holds "that consciousness only arises in a late derivative phase of complex integrations. If an actual occasion be such that phases of this sort are negligible in its concrescence, then in its experience there is no knowledge; owing to the fact that consciousness is a subjective form belonging to the later phases, the prehensions which it directly irradiates are those of an 'impure' type. Consciousness only illuminates the more primitive types of prehensions so far as these prehensions are still elements in the products of integration. Thus those elements of our experience which stand out clearly and distinctly in our consciousness are not its basic facts; they are the derivative modifications which arise in the process. For example, consciousness only dimly illuminates the prehensions in the mode of causal efficacy, because these prehensions are primitive elements in our experience. But prehensions in the mode of presentational immediacy are among those prehensions which we enjoy with the most vivid consciousness. These prehensions are late derivatives in the concrescence of an experient subject. The consequences of the neglect of this law, that the late derivative elements are more clearly illuminated by consciousness than the primitive elements, have been fatal to the proper analysis of an experient occasion. In fact, most of the difficulties of philosophy are produced by it" [PR 245–246]. (See also SYMBOLIC REFERENCE.)

Contemporaneousness—Contemporary actual entities are actual entities that occur in causal independence of one another—i.e., none of which are objectified as a datum for the initial, receptive phase of the concrescence of any of the others. "Actual entities are called 'contemporary' when neither belongs to the 'given' actual world defined by the other" [PR 102].

Einstein's relativity theory is incorporated into Whitehead's account of contemporaneousness. Given the above definition it is quite possible for a certain actual entity M to be a contemporary of actual entity P and also of actual entity N, and yet for P and N not to be contemporaries. (See Figure 4 and accompanying text of this book for a fuller account; see also DURATION.)

Contemporary regions are known by means of perception in the mode of presentational immediacy. This mode of perception provides "clear, distinct consciousness of the 'extensive' relations of the world" [PR 95]. Contemporary actual entities do not enter into the constitution of one another by objectification of each other's feelings; their only connection "is their implication in the same extensive scheme" [*PR* 484]. This connection is a result of the fact that two contemporary actual entities, *A* and *B*, "are atomic regions in the potential scheme of spatio-temporal extensiveness which is a datum for both *A* and *B*" [*PR* 188]. Both *A* and *B* occur in a chain of extensive connectedness such that, given the past weight of spatial, temporal, and geometrical relationships, the extension of these modes of extensive connectedness is the only real potentiality open to the unfolding universe in their particular cosmic epoch. Hence both *A* and *B* feel something about the extensive properties of each other, because each feels directly the past out of which they both arise. It is this awareness of extensive properties that is revealed about contemporaries in perception in the mode of presentational immediacy. (*See also* PRESENTATIONAL IMMEDIACY.)

Contrast—A contrast is the unity had by the many components in a complex datum. The name is somewhat misleading, for *to set in contrast with* means *to put in a unity with. Contrast* is the opposite of *incompatibility.* The more an actual entity can hold the items of its experience in contrasts, and contrasts of contrasts, the more it elicits depth and intensity for its satisfaction. Primitive actual entities that are unable to hold the items of their experience in contrasts are forced, because of the resulting incompatibilities, to dismiss some items as irrelevant, with the result that their experience is relatively shallow and trivial.

Corpuscular society—A corpuscular society is a special kind of structured society. (*For background, see* NEXUS, SOCIETY, *and* STRUCTURED SOCIETY.) The special characteristic of a corpuscular society is that the subordinate societies constitutive of it are all strands of enduring objects. An enduring object is a society that is purely temporal and continuous—i.e., it is a mere thread of continuous inheritance containing no two actual entities that are contemporaries. "A society may be more or less corpuscular, according to the relative importance of the defining characteristics of the various enduring objects compared to that of the defining characteristic of the whole corpuscular nexus" [*PR* 52].

Cosmic epoch—The properly metaphysical characteristics of actual entities are those characteristics shared by all actual entities. There are, however, many characteristics of any given set of actual entities that are not properly metaphysical, but are the result of special, less than universal, social relations. The notion of our cosmic epoch is the notion of a vast society establishing relations that are admittedly not universal, not perfectly general, but which constitute "that widest society of actual entities whose immediate relevance to ourselves is traceable. This epoch is characterized by electronic and protonic actual entities, and by yet more ultimate actual entities which can be dimly discerned in the quanta of energy" [*PR* 139].

One interesting implication of this doctrine is that Whitehead holds that the laws of nature evolve. "But there is not any perfect attainment of an ideal order whereby the indefinite endurance of a society is secured. A society arises from disorder, where 'disorder' is defined by reference to the ideal for that society; the favourable background of a larger environment either itself decays, or ceases to favour the persistence of the society after some stage of growth: the society then ceases to reproduce its members, and finally after a stage of decay passes out of existence. Thus a system of 'laws' determining reproduction in some portion of the universe gradually rises into dominance; it has its stage of endurance, and passes out of existence with the decay of the society from which it emanates" [*PR* 139]. Our cosmic epoch is constituted by some such system of laws and upon an unimaginable lapse of time will have given way to new, presently inconceivable forms of social order.

The result of this analysis is to make the laws of nature much less mysterious and miraculous than some other contemporary accounts make them. "Maxwell's equations of the electromagnetic field hold sway by reason of the throngs of electrons and protons. Also each electron is a society of electronic occasions, and each proton is a society of protonic occasions. These occasions are the reasons for the electromagnetic laws . . . . But there is disorder in the sense that the laws are not perfectly obeyed, and that the reproduction is mingled with instances of failure. There is accordingly a gradual transition to new types of order, supervening upon a gradual rise into dominance on the part of the present natural laws" [*PR* 139–140].

Creativity—*Creativity* is one of three notions involved in what Whitehead calls the Category of the Ultimate; this category expresses the general principle presupposed by all other aspects of the philosophy of organism (Whitehead's name for his own position). The other two notions involved are *many* and *one*.

Whitehead's philosophy is a process philosophy, and the notion of creativity is crucial to an understanding of process. The basic presupposition of the whole system is ongoingness: generation after generation of actual entities succeeding one another without end. Creativity expresses that ultimate fact about actual entities that makes ongoingness intelligible.

The principle of creativity enunciates the following relationships between *many* and *one*: (1) at any instant the universe constitutes a disjunctively diverse *many*; (2) "it lies in the nature of things that the many enter into complex unity" [*PR* 31]; (3) the novel *one* that results from this unification, this concrescence, is truly novel—i.e., it stands over and against what has been unified and as such is disjunctively diverse from the items it has unified; and (4) there is here the same situation from which the process began (i.e., a disjunctive diversity) and it therefore repeats itself "to the crack of doom in the creative advance from creature to creature" [*PR* 347]. "The ultimate metaphysical principle is the advance from disjunction to conjunction, creating a novel entity other than the entities given in disjunction" [*PR* 32].

Whitehead's understanding of creativity does not do violence to the ontological principle; creativity is not, nor does it point to, some kind of entity or being more real than actual entities. It is, rather, descriptive of the most fundamental relationships participated in by all actual entities. " 'Creativity' is the universal of universals characterizing ultimate matter of fact" [*PR* 31]. (*See also* PROCESS.)

Datum—The datum of a prehension is that which is prehended, it is what is given. "In the analysis of a feeling, whatever presents itself as also *ante rem* is a datum . . ." [*PR* 355]. "The character of an actual entity is finally governed by its datum; whatever be the freedom of feeling arising in the concrescence, there can be no transgression of the limitations of capacity inherent in the datum. The datum both limits and supplies. It follows from this doctrine that the character of an organism depends on that of its environment" [*PR* 168].

Whitehead distinguishes an *initial datum* from an *objective datum*. This distinction is relevant when simple physical feelings are being considered—i.e., feelings that have as datum just one actual entity. The entire actual entity that is the datum is termed the *initial datum*. That initial datum is objectified for the concrescing subject by one of its component feelings. That objectifying component feeling is termed the *objective datum*. "Objectification relegates into irrelevance, or into a subordinate relevance, the full constitution of the objectified entity. Some real component in the objectified entity assumes the rôle of being how that particular entity is a datum in the experience of the subject." [*PR* 97]. This "real component" is the objective datum, and the objectification it effects is termed the *perspective* of the initial datum for the concrescing subject.

Defining characteristic—*See* Society.

Duration—A duration is a cross section of the universe "defined by the characteristic that any two of its members are contemporaries" [*PR* 192]. Given any actual entity, *M,* any member of a duration including *M* is a contemporary of *M.* "The characteristic property of a duration is termed 'unison of becoming'" [*PR* 192].

The classical theory of time assumed that there was but one duration through any actual entity. Whitehead's mode of incorporating the modern, relativity theory of time into his system is to assert that there is more than one duration through a given actual entity. Every member of every duration through *M* will be a contemporary of *M,* but because there is more than one duration through *M,* it follows that there may be two occasions, both contemporaries of *M,* which are not themselves contemporaries, being members of different durations. This is an instance of an effect of what the new theory refers to as the warping of space and time.

But Whitehead recognizes that the overwhelming obviousness of conviction that accompanies the ordinary belief in the notion of "the immediate present condition of the world at some epoch" [*PR* 190] entails that "some measure of acceptance is imposed upon metaphysics. If the notion be wholly rejected no appeal to universal obviousness of conviction can have any weight; since there can be no stronger instance of this force of obviousness" [*PR* 191]. Whitehead's solution to the conflict between obviousness of conviction and recent physics is to isolate one particular duration associated with any given actual entity and make this duration answer to ordinary convictions. This particular duration is called the *presented duration,* and it is a function of its associated entity's perception in the mode of presentational immediacy. In this mode of perception sensa illustrate potential subdivisions within a cross section of the world; this cross section is that entity's immediate present, and that entity's presented duration is the one duration that includes all of its immediate present.

Efficient cause—*See* Causal efficacy.

Enduring object—"An 'enduring object,' or 'enduring creature,' is a society whose social order has taken the special form of 'personal order' " [*PR* 50]. A personally ordered society is ordered serially —i.e., it is a purely temporal society, a mere thread of continuous inheritance containing no two actual entities that are contemporaries. (*For a fuller account see* NEXUS, SOCIETY, CORPUSCULAR SOCIETY, *and* STRUCTURED SOCIETY.)

Eternal object—"Any entity whose conceptual recognition does not involve a necessary reference to any definite actual entities of the temporal world is called an 'eternal object' " [*PR* 70]. Eternal objects are forms of definiteness capable of specifying the character of actual entities; they are "Pure Potentials for the Specific Determination of Fact" [*PR* 32]. An actual entity's process of becoming is a process of acquiring definiteness by a series of decisions to select or reject various forms of definiteness (eternal objects). "The determinate definiteness of each actuality is an expression of a selection from these forms. It grades them in a diversity of relevance" [*PR* 69]. It is the actual entities that do the selecting and rejecting: "an eternal object is always a potentiality for actual entities; but in itself, as conceptually felt, it is neutral as to the fact of its physical ingression in any particular actual entity of the temporal world" [*PR* 70].

Any given actual entity does not make its decisions with utter freedom. "An actual entity arises from decisions *for* it and by its very existence provides decisions *for* other actual entities which supersede it" [*PR* 68]. The past, from which it inherits, presents it with certain forms of definiteness that it is compelled to reiterate. "Some conformation is necessary as a basis of vector transition, whereby the past is synthesized with the present. The one eternal object in its two-way function, as a determinant of the datum and as a determinant of the subjective form, is thus relational. . . . An eternal object when it has ingression through its function of objectifying the actual world, so as to present the datum for prehension, is functioning 'datively' " [*PR* 249].

To be an individual, concrete fact each actual entity must assume some determinate form; this it does by means of its decisions as to which eternal objects it will permit, and which eternal objects it will not permit, to become ingredient in its concrescence. The process, the becoming involved in the decision of an actual entity is one with its very being; eternal objects, on the other hand, are essentially aloof from change in that it is of their essence to be eternal. But they are involved in change in the sense that the very process of becoming that is any given actual occasion is the process of determining, via selected eternal objects, the specific character, the kind of definiteness, that will make that actual entity what it will be. "The actualities constituting the process of the world are conceived as exemplifying the ingression (or 'participation') of other things which constitute the potentialities of definiteness for any actual existence. The things which are temporal arise by their participation in the things which are eternal" [*PR* 63].

This sounds Platonic, and to a degree it is meant to be.

Whitehead distinguishes between eternal objects of the subjective species and eternal objects of the objective species, and explicitly states that "eternal objects of the objective species are the mathematical platonic forms" [*PR* 446]. These forms are objective in the sense that they are "an element in the definiteness of some objectified nexus, or of some single actual entity which is a datum of a feeling" [*PR* 445]. An eternal object of the subjective species is subjective in the sense that it is "an element in the definiteness of the subjective form of a feeling. . . . It is an emotion, or an intensity, or an adversion, or an aversion, or a pleasure, or a pain" [*PR* 446]. Whitehead's scheme is *not* Platonic in that it does not allow an eminent reality to the realm of eternal objects. The ontological principle assigns to actual entities reality in the fullest sense of the term, and here Whitehead embodies Aristotle's protest against Plato's "other worldliness." But, by the ontological principle, "everything must be somewhere; and here 'somewhere' means 'some actual entity.' Accordingly the general potentiality of the universe [i.e., the realm of eternal objects] must be somewhere; since it retains its proximate relevance to actual entities for which it is unrealized. . . . This 'somewhere' is the non-temporal actual entity. Thus 'proximate relevance' means 'relevance as in the primordial mind of God'" [*PR* 73]. (*See* GOD *for a further discussion.*)

Event—In Whitehead's earlier works the notion of an event was central; events played a role analogous to that played by actual entities in the later works (which include *Process and Reality*). Events could be temporally quite extended in the earlier works. But an actual entity occurs in an instant; its becoming is also its moment of perishing. So in this respect events and actual entities are quite different.

Whitehead retains the notion of an event in the later works, but it ceases to be as fundamental a term as it was earlier. In *Process and Reality* an event "is a nexus of actual occasions interrelated in some determinate fashion in some extensive quantum: it is either a nexus in its formal completeness, or it is an objectified nexus. . . . An actual occasion is the limiting type of an event with only one member. . . . For example, a molecule is a

historic route of actual occasions; and such a route is an 'event' "
[*PR* 124, 113, 124].

Extensive continuum—Societies do not exist in isolation; each presupposes its social environment, and they nestle inside one another as a town lies in a county, a county in a state, a state in a
nation. Of course, just as there are many counties in a state, so
there may be many societies of a type within one containing society. Nature is composed, then, of "a series of societies of increasing width of prevalence, the more special societies being included
in the wider societies" [*PR* 141]. Our cosmic epoch (*which see*) is
dominated by a vast society of electronic and protonic actual
entities, and the order exhibited by this society is what we call
the *laws of Nature*. But this society is set in a wider social context that exhibits the geometrical axioms, and within this wider
society there might be an antielectronic, antiprotonic society, just
as there may be two towns within a county. The geometrical
society, however, presupposes an even wider society of four-
dimensionality, and beyond that there is distinguishable a society
of mere dimensionality, and finally, the widest society conceivable
in our present state of knowledge, a society of mere extensiveness,
which is called by Whitehead the *extensive continuum*. "In these
general properties of extensive connection, we discern the defining characteristic of a vast nexus extending far beyond our immediate cosmic epoch. . . . This ultimate, vast society constitutes
the whole environment within which our epoch is set, so far as
systematic characteristics are discernible by us in our present stage
of development" [*PR* 148].

The concept of the extensive continuum has now been arrived
at by working outwards in terms of societies; it is illuminating
also to approach the concept working inward from the notion of
pure, or general, potentiality. The general potentiality "is the
bundle of possibilities, mutually consistent or alternative, provided by the multiplicity of eternal objects" [*PR* 102]; it is the
realm of eternal objects considered in itself apart from any limitations, other than logical ones, that might be put upon ingression into the world. But there *are*, in fact, limitations put upon
ingression into the world, not only by logic, but by past circum-

stances—in terms of pure possibility it is possible that a five-foot, fortyish, fat man could high jump seven feet, but this is not a real possibility. Real potentiality is the limitations, the restrictions upon pure potentiality, that the conditions of a given, factual world impose upon any particular actual entity arising out of that world. The extensive continuum "is that first determination of order—that is, of real potentiality—arising out of the general character of the world. In its full generality beyond the present epoch, it does not involve shapes, dimensions, or measurability; these are additional determinations of real potentiality arising from our cosmic epoch" [*PR* 103]. As a vast society reaching beyond our cosmic epoch, the extensive continuum lays down, through the massive social inheritance of its myriad generations, the first, most general limitation upon general potentiality, the limitation that each generation of actual entities, no matter what its more special characteristics of order, shall at least exhibit the general properties "of 'extensive connection,' of 'whole and part,' of various types of 'geometrical elements' derivable by 'extensive abstraction'; but excluding the introduction of more special properties by which straight lines are definable and measurability thereby introduced" [*PR* 148]. The technical geometrical aspects of this theory are beyond this introductory study; suffice it to say that the space-time continuum is a set of more special characteristics that presuppose mere extension and that "extension, apart from its spatialization and temporalization, is that general scheme of relationships providing the capacity that many objects can be welded into the real unity of one experience. Thus, an act of experience has an objective scheme of extensive order by reason of the double fact that its perspective standpoint has extensive content, and that the other actual entities are objectified with the retention of their extensive relationships" [*PR* 105].

Feeling—A feeling is a positive prehension (*see* PREHENSION). "This word 'feeling' is a mere technical term; but it has been chosen to suggest that functioning through which the concrescent actuality appropriates the datum so as to make it its own" [*PR* 249]. Whitehead makes much of certain passages in Locke where Locke speaks of "ideas determined to particulars." The point here is that, at some moments at least, Locke deserts a representative theory of perception and holds that in experience our ideas are not mere representations, but vehicles whereby subjects grasp other entities and incorporate them as components in their own constitutions. The technical term *feeling* is an effort to incorporate this occasional usage on the part of Locke into the heart of the philosophy of organism. "This notion of a direct 'idea' (or 'feeling') of an actual entity is a presupposition of all common sense. . . . Each actual entity is conceived as an act of experience arising out of data. It is a process of 'feeling' the many data, so as to absorb them into the unity of one individual 'satisfaction.' Here 'feeling' is the term used for the basic generic operation of passing from the objectivity of the data to the subjectivity of the actual entity in question" [*PR* 83, 65].

Formaliter—An actual entity considered *formaliter* is an actual entity considered as subjective, as enjoying its own immediacy of becoming. "I will adopt the pre-Kantian phraseology, and say that the experience enjoyed by an actual entity is that entity *formaliter*. By this I mean that the entity, when considered 'formally,' is being described in respect to those forms of its constitution whereby it is that individual entity with its own measure of absolute self-realization. . . . The 'formal' reality of the actuality in question belongs to its process of concrescence and not to its 'satisfaction' " [*PR* 81, 129]. (*See also* OBJECTIVÉ.)

God—"God is an actual entity, and so is the most trivial puff of existence in far-off empty space. But, though there are gradations of importance, and diversities of function, yet in the principles which actuality exemplifies all are on the same level. . . . The presumption that there is only one genus of actual entities constitutes an ideal of cosmological theory to which the philosophy of organism endeavours to conform" [*PR* 28, 168]. There is difference of opinion among experts as to exactly how well Whitehead's account of God conforms to this ideal, but it is clear that Whitehead intends that in his philosophy "God is not to be treated as an exception to all metaphysical principles, invoked to save their collapse. He is their chief exemplification" [*PR* 521].

The structure of God is the mirror image to structure in the world. The world is incomplete; in its very nature it requires an entity at the base of all things to complete it. This entity is God, and as the entity that completes the world God is the complement of the world—as the left hand is the complement of the right—with the result that the principles governing all actual entities are in some instances exemplified in a reverse way in God.

The entities of the temporal world originate with *physical* prehensions of datum occasions—this roots them firmly in space and time. God originates with his *conceptual* valuation of the timeless realm of eternal objects—this is the *primordial nature* of God and is the basis for referring to God as the nontemporal actual entity. Actual entities of the temporal realm proceed from physical prehensions to conceptual prehensions; God reverses this, for his *consequent nature* is constituted by his physical prehensions of the actual entities in the temporal world. God, like temporal occasions, is completed by phases of comparative feelings in which the physical feelings of his consequent nature are integrated with the conceptual feelings of his primordial nature. As with temporal entities, in God this integration also produces a *superjective nature,* which is "the character of the pragmatic value of his specific satisfaction qualifying the transcendent creativity in the various temporal instances" [*PR* 135].

Eternal objects, considered as a realm, constitute a kind of existence that, by the ontological principle, requires its link with actuality. "Everything must be somewhere; and here 'somewhere' means 'some actual entity.' Accordingly the general potentiality of the universe must be somewhere . . . . This 'somewhere' is the nontemporal actual entity . . . [i.e.,] the primordial mind of God" [*PR* 73]. This is one way in which the internal structure of the system requires God.

The system also requires God in connection with the doctrine of objectification. Actual entities become and then perish, but they also function as objectively immortal. As objectively immortal, actual entities have lost their link with actuality so that in this respect they are like the realm of eternal objects. Objectively immortal actual entities find their link with actuality through the consequent nature of God; considered as consequent, God prehends and preserves each generation of actual entities and mediates their efficacy in the future. In this role God exhibits his traditional saving power and his judging power. God's consequent nature grows with the growth of the world, and through his valuations of the world as saved in his consequent nature he exhibits "the judgment of a tenderness which loses nothing that can be saved" [PR 525].

Finally, the system requires an entity that mediates between actuality and potentiality in such a way that novelty and progress are possible. In his primordial nature God prehends the infinite realm of possibilities; in his consequent nature he prehends the actualities of the world; his superjective nature is a result of weaving his consequent prehensions upon his primordial vision. As actual fact is thereby brought into juxtaposition with the realm of possibilities, relevant, but novel, possibilities for that factual situation emerge into important contrast with what has in fact occurred. God's own aim in the creative advance is to have a world emerge of such a sort that his own experience of that world will result in the greatest possible intensity in his own experience. He therefore—and this is God functioning superjectively—offers as a lure to each actual entity as it arises that subjective aim the completion of which, in that entity's own concrescence, would create the kind of ordered, complex world that, when prehended by God, would result in maximum intensity of satisfaction for him. In this limited sense of providing its initial subjective aim, "God can be termed the creator of each temporal actual entity" [PR 343]. (See also SUBJECTIVE AIM.)

Hybrid prehension—See Prehension.

Impure prehension—See Prehension.

Ingression—This term approximates to Plato's notion of *participation;* it indicates the manner in which eternal objects are present in actual entities. (See also ETERNAL OBJECT.)

Intellectual feeling—Intellectual feelings are the complex comparative feelings that constitute Phase IV of any given sophisticated concrescence (*see Figure 2 of this book*). They feel the contrast between theory and fact, between what might be the case and what is. The elements contrasted are a proposition—a theory—from Phase III and the feeling of the fact from Phase I that contains the logical subjects of that proposition. Consciousness arises as the subjective form of the feeling of this contrast, sometimes referred to as the *affirmation-negation contrast*. "If some eternal objects, in their abstract capacity, are realized as relevant to actual fact, there is an actual occasion with intellectual operations" [*PR* 326].

Life—*See* Structured society.

Logical subjects (of a proposition)—*See* Proposition.

Mental pole—For some purposes of analysis Whitehead breaks an actual entity into two distinguishable parts, its mental pole and its physical pole. The physical pole answers to what, in Figure 2 of this book, is labeled the first phase of concrescence, the initial phase of conformal feelings; the mental pole answers to what in Figure 2 is labeled the supplemental phases—i.e., the originative phases of conceptual feelings and comparative feelings.

The physical pole is that aspect of an actual entity wherein it makes no contribution of its own, but merely receives what is given for it from the past. The mental pole is that aspect of an actual entity which responds to what is given—"The mental pole introduces the subject as a determinant of its own concrescence. The mental pole is the subject determining its own ideal of itself . . . . The dipolar character of concrescent experience provides in the physical pole for the objective side of experience, derivative from an external actual world, and provides in the mental pole for the subjective side of experience, derivative from the subjective conceptual valuations correlate to the physical feelings" [*PR* 380, 423].

The terms *physical pole* and *mental pole* may not be the happiest of terms to introduce into a philosophy that repudiates the

Cartesian dualism and insists that actual entities are the only finally real actualities. Certainly Whitehead has no intention of reintroducing the old concepts of mind and matter, and it is emphatically *not* the case that actual entities in the physical world have only physical poles and that mental poles are present only in the higher organisms. "No actual entity is devoid of either pole; though their relative importance differs in different actual entities. . . . Thus an actual entity is essentially dipolar, with its physical and mental poles; and even the physical world cannot be properly understood without reference to its other side, which is the complex of mental operations" [*PR* 366]. On the other hand, this does not mean "that these mental operations involve consciousness, which is the product of intricate integration" [*PR* 379].

Metaphysics—Metaphysics, or Speculative Philosophy, "is the endeavour to frame a coherent, logical, necessary system of general ideas in terms of which every element of our experience can be interpreted. By this notion of 'interpretation' I mean that everything of which we are conscious, as enjoyed, perceived, willed, or thought, shall have the character of a particular instance of the general scheme. . . . The true method of philosophical construction is to frame a scheme of ideas, the best that one can, and unflinchingly to explore the interpretation of experience in terms of that scheme" [*PR* 4, x].

Whitehead uses the adjective *metaphysical* when discussing characteristics of actual entities that are completely general. "The metaphysical characteristics of an actual entity—in the proper general sense of 'metaphysics'—should be those which apply to all actual entities" [*PR* 138]. There is no hint of dogmatism in Whitehead's attitude: "It may be doubted whether such metaphysical concepts have ever been formulated in their strict purity —even taking into account the most general principles of logic and of mathematics. We have to confine ourselves to societies sufficiently wide, and yet such that their defining characteristics cannot safely be ascribed to all actual entities which have been or may be. . . . In philosophical discussion, the merest hint of dogmatic certainty as to finality of statement is an exhibition of folly" [*PR* 138–139, x].

Multiplicity—A multiplicity is what answers in Whitehead's system to the notion of a logical class. "A multiplicity consists of many entities, and its unity is constituted by the fact that all its constituent entities severally satisfy at least one condition which no other entity satisfies" [*PR* 36]. A multiplicity is to be contrasted with a society, which has its unity as a result of a genetically sustained ordering characteristic. As a result a society is self-sustaining, whereas a multiplicity involves only a mathematical conception of order and is simply a set of entities to which the same class-name applies. "A multiplicity is a type of complex thing which has the unity derivative from some qualification which participates in each of its components severally; but a multiplicity has no unity derivative *merely* from its various components" [*PR* 73].

Nexus—An actual entity is a microcosmic entity; the macrocosmic entities of everyday experience—men, trees, houses—are groupings of entities termed nexūs (plural of nexus), or societies. Though for most purposes the terms *society* and *nexus* are interchangeable, the class of nexūs is wider than the class of societies; all societies are nexūs, but not all nexūs are societies.

"Actual entities involve each other by reason of their prehensions of each other. There are thus real individual facts of the togetherness of actual entities, which are real, individual, and particular, in the same sense in which actual entities and the prehensions are real, individual, and particular. Any such particular fact of togetherness among actual entities is called a 'nexus'" [*PR* 29–30].

Mutual immanence is the most general common function of the actual entities constituting a nexus. The most common condition is for a nexus to spread itself both spatially and temporally. A tree, which at an instant is many actual entities thick spatially, is also many generations of actual entities thick temporally. There are two limiting kinds of nexūs, one that is purely temporal and one that is purely spatial. A purely temporal nexus contains no pair of contemporary actual occasions; it is a mere thread of temporal transition from occasion to occasion, and the mutual immanence involved is the causal immanence of each actual entity prehending the entity immediately preceding it in the thread. A purely spatial nexus includes no pair of occasions such that one of the pair is antecedent to the other; it is a slice through time composed of contemporary actual occasions, and the mutual immanence involved is of the indirect type proper to contemporary occasions—i.e., it results from mutual implication in one scheme of extensive connectedness. (*See* CONTEMPORANE-OUSNESS *for a fuller account.*)

A society is a nexus that enjoys social order—i.e., one that exhibits characteristics in each generation of actual entities that are derived from prehensions of previous generations. It follows that a purely spatial nexus cannot be a society. The concept of a society is associated with the notion of order; this leads White-head to assert that "A non-social nexus is what answers to the notion of 'chaos' " [*PR* 112].

Nexūs and societies admit of great complexity; there are struc-tured societies—i.e., societies that include subordinate societies and/or subordinate nexūs. The difference between a subordinate society and a subordinate nexus within a structured society is that a subordinate society can retain its dominant features in the general environment apart from the structured society, whereas a subordinate nexus presents "no features capable of genetically sustaining themselves apart from the special environment pro-vided by that structured society" [*PR* 151–152].

Objectification—"The discussion of how the actual particular occa-sions become original elements for a new creation is termed the theory of objectification. . . . The functioning of one actual entity in the self-creation of another actual entity is the 'objectification' of the former for the latter actual entity" [*PR* 320–321, 38].

The theory of objectification is central to Whitehead's episte-mology. As embodied in the Principle of Relativity, the theory of objectification seeks to overthrow the basic assumptions that lead ultimately to Hume's skepticism. The Principle of Relativity states "that the potentiality for being an element in a real concrescence of many entities into one actuality, is the one general metaphysical character attaching to all entities, actual and nonactual" [*PR* 33]. This means that each actual occasion, once it has become and reached its satisfaction, loses its subjectivity, its own immediacy of becoming, and serves as a datum for succeeding generations of actual entities, which incorporate it, in some aspect, into their very being by prehending it as a datum to be absorbed into their own concrescences. The subject-predicate, substance-quality, particular-universal dichotomies, which stem from Aristotle and lead to Hume, are all repudiated by this theory of objectification. "The principle of universal relativity directly traverses Aristotle's dictum, '(A substance) is not present in a subject.' On the contrary, according to this principle an actual entity *is* present in other actual entities. . . . The philosophy of organism is mainly devoted to the task of making clear the notion of 'being present in another entity.' This phrase is here borrowed from Aristotle: it is not a fortunate phrase, and in subsequent discussion it will be replaced by the term 'objectification' " [*PR* 79–80].

Whitehead's doctrine of objectification is that other actual entities *are* in a given actual entity, *a*, in the sense that their relationships with *a* are constitutive of the essence of *a*. The prehensions of *a*—i.e., *a*'s prehensions—constitute both its relationships to these other entities and its own essence. The role of eternal objects is central in this objectification. "The Aristotelian phrase suggests the crude notion that one actual entity is added to another *simpliciter*. This is not what is meant. One rôle of the eternal objects is that they are those elements which express how any one actual entity is constituted by its synthesis of other actual entities . . . . The eternal objects function by introducing the multiplicity of [datum] actual entities as constitutive of the [concrescing] actual entity in question. . . . This is the direct denial of the Cartesian doctrine, '. . . an existent thing which requires nothing but itself in order to exist' " [*PR* 80, 93]. (*See* ETERNAL

OBJECT *for a further account of eternal objects functioning da-*
tively; see also PREHENSION.)

Objectivé—"An 'object' is a transcendent element characterizing
that *definiteness* to which our 'experience' has to conform" [*PR*
327] and an actual entity considered *objectivé* is an actual entity
considered *not* as *formaliter* (*which see*), *not* as enjoying its own
immediacy of becoming, but rather as dead datum that has become brute fact and that consequently, as objectively immortal,
conditions all concrescence beyond itself as something given, as
an object. "In Descartes' phraseology, the satisfaction is the actual
entity considered as analysable in respect to its existence '*objectivé*.' It is the actual entity as a definite, determinate, settled fact,
stubborn and with unavoidable consequences" [*PR* 335–336]. (*See
also* SATISFACTION, SUPERJECT, *and* OBJECTIVE IMMORTALITY.)

Objective immortality—"The attainment of a peculiar definiteness
is the final cause which animates a particular process; and its attainment halts its process, so that by transcendence it passes into
its objective immortality as a new objective condition added to
the riches of definiteness attainable, the 'real potentiality' of the
universe" [*PR* 340]. (*See also* OBJECTIVÉ, SUPERJECT, *and* SATISFAC-
TION.)

Ontological principle—The ontological principle asserts that "every
condition to which the process of becoming conforms in any
particular instance, has its reason *either* in the character of some
actual entity in the actual world of that concrescence, *or* in the
character of the subject which is in process of concrescence. . . .
According to the ontological principle there is nothing which
floats into the world from nowhere" [*PR* 36, 373]. God is a part
of the actual world of every actual occasion, and he includes in
his primordial nature the realm of eternal objects. "By this recognition of the divine element the general Aristotelian principle is
maintained that, apart from things that are actual, there is
nothing—nothing either in fact or in efficacy. . . . Thus the
actual world is built up of actual occasions; and by the ontologi-

cal principle whatever things there are in any sense of 'existence,' are derived by abstraction from actual occasions" [*PR* 64, 113].

Personally ordered society—*See* Enduring object.

Physical pole—*See* Mental pole.

Physical prehension—*See* Conformal feeling.

Physical purpose—Physical purposes are the most primitive kind of simple comparative feelings that can arise in Phase III of concrescence (*see Figure 2 of this book*). In a simple comparative feeling there is the integration of a simple physical feeling (from Phase I) with its conceptual counterpart—i.e., the conceptual feeling derived from it (in Phase II). It is the character of this integration that distinguishes physical purposes from the more sophisticated propositional feelings (*which see*). The difference lies in the fate of the conceptual feeling from Phase II. If in the integration the eternal object that is the datum of the conceptual feeling retains its indetermination to fact, its universality, its transcendence of fact, then the integration results in a propositional feeling at Phase III. If, however, the eternal object *loses* its indetermination to fact, its transcendence, and sinks back into immanence in fact, into union with itself as originally exemplified in the physical feeling in Phase I, then the integration results in a physical purpose at Phase III.

A physical purpose is for all practical purposes a reiteration of the physical feeling felt in Phase I, except that the subjective form of the conceptual feeling in Phase II may be either adversion or aversion. If the former, then the actual entity that is the subject of the physical purpose tends to preserve that physical feeling and transmit it to future occasions; if the latter, then in some way the physical feeling will tend to lose importance in the future beyond that subject.

Whereas propositional feelings are lures for more sophisticated integrations and integrations of integrations that can result in consciousness, physical purposes tend to be terminal—they inhibit further integration, are devoid of consciousness, and characterize the primitive sorts of actual entities that are members of the kinds of societies we term inanimate objects.

Potentiality; general and real—*See* Extensive continuum.

Predicative pattern—*See* Proposition.

Prehension—Prehensions are defined as "Concrete Facts of Relatedness" [PR 32]. Prehensions are the vehicles by which one actual entity becomes objectified in another, or eternal objects obtain ingression into actual entities; they "are 'vectors'; for they feel what is *there* and transform it into what is *here*" [PR 133].

Prehensions are what an actual entity is composed of: "The first analysis of an actual entity, into its most concrete elements, discloses it to be a concrescence of prehensions, which have originated in its process of becoming" [PR 35]. The very nature of a prehension reveals its relational character: "Every prehension consists of three factors: (a) the 'subject' which is prehending, namely, the actual entity in which that prehension is a concrete element; (b) the 'datum' which is prehended; (c) the 'subjective form' which is *how* that subject prehends that datum" [PR 35].

Physical prehensions are prehensions whose data involve actual entities; *conceptual prehensions* are prehensions whose data involve eternal objects. Both physical and conceptual prehensions are spoken of as *pure*; an *impure prehension* is a prehension in a later phase of concrescence that integrates prehensions of the two pure types. A *hybrid prehension* is the "prehension by one subject of a conceptual prehension, or of an 'impure' prehension, belonging to the mentality of another subject" [PR 163]. A *positive prehension* (also termed a *feeling*) includes its datum as part of the synthesis of the subject occasion, but *negative prehensions* exclude their data from the synthesis.

"The perceptive constitution of the actual entity presents the problem, How can the other actual entities, each with its own formal existence, also enter objectively into the perceptive constitution of the actual entity in question? This is the problem of the solidarity of the universe. The classical doctrines of universals and particulars, of subject and predicate, of individual substances not present in other individual substances, of the externality of relations, alike render this problem incapable of solution. The answer given by the organic philosophy is the doctrine of prehensions, involved in concrescent integrations, and terminating in a definite, complex unity of feeling" [*PR* 88–89].

Whitehead acknowledges an indirect debt to Leibniz in his use of this term. Leibniz employed the terms *perception* and *apperception* for the lower and higher ways, respectively, that one monad can take account of another, can be aware of another. While needing a set of terms like this, Whitehead does not wish to utilize the identical terminology, for as used by Leibniz the terms are inextricably bound up with the notion of representative perception, which Whitehead rejects. But there is the similar term *apprehension,* meaning "thorough understanding," and, using the Leibnizian model, Whitehead coins the term *prehension* to mean the *general, lower* way, devoid of any suggestion of either consciousness or representative perception, in which an occasion can include other actual entities, or eternal objects, as part of its own essence. (*For a further discussion see* OBJECTIFICATION *and* FEELING.)

Presentational immediacy—Presentational immediacy is the more sophisticated and complex of the two pure modes of perception. (Causal efficacy is the other pure mode, and both combine to form the mixed mode of perception, symbolic reference, which is our ordinary mode of awareness.)

Presentational immediacy is the perceptive mode "in which there is clear, distinct consciousness of the 'extensive' relations of the world. . . . In this 'mode' the contemporary world is consciously prehended as a continuum of extensive relations" [*PR* 95]. Whereas causal efficacy, the mode of inheritance from the past, transmits, into the present, data that are massive in emotional power but vague and inarticulate, presentational immediacy transmits data that are sharp, precise, spatially located, but isolated, cut off, self-contained temporally; there is no power of continuity to them for they are simply an awareness of those extensive relationships that constitute the contemporary world for the prehending subject.

Presentational immediacy is an elaboration upon certain aspects of what is present already in causal efficacy—this is possible because although perception in the mode of causal efficacy occurs in the first phase of concrescence, perception in the mode of presentational immediacy occurs in later phases and presupposes causal efficacy. In particular, causal efficacy contains sensa, but in a vague, ill-defined, and hardly relevant way. Presentational immediacy seizes upon these vague emotional feelings and transforms them into sharp qualities that are then projected into the contemporary region of that percipient occasion. The result is a flashing awareness of, say, "gray, there," which is a typical example of perception in the mode of presentational immediacy. This is the kind of awareness one has when a flick of color is noted out of the corner of an eye. It is not until perception in the mixed mode is attained that there is the ordinary awareness of the persisting gray stone. (*See also* CAUSAL EFFICACY, SYMBOLIC REFERENCE, *and* CONTEMPORANEOUSNESS.)

Presented locus—The presented locus of any given actual entity is that entity's "contemporary nexus perceived in the mode of presentational immediacy, with its regions defined by sensa" [*PR* 192]. Whitehead sometimes refers to the presented locus as the *immediate present* of an actual entity, or its *presented duration*. (*See also* DURATION, *where this concept is discussed under the name* presented duration.)

Primary feeling—"In two extreme cases the initial data of a feeling have a unity of their own. In one case, the data reduce to a single actual entity, other than the subject of the feeling; and in the other case the data reduce to a single eternal object. These are called 'primary feelings' " [*PR* 353]. Whitehead also refers to primary feelings as *simple feelings*. He discusses *simple physical feelings* at some length—these would be primary feelings of the first species. Such simple physical feelings are *conformal feelings* (*which see*) though not all conformal feelings need be simple feelings.

Process—"There are two kinds of fluency. . . . One kind is the fluency inherent in the constitution of the particular existent. This kind I have called 'concrescence.' The other kind is the fluency whereby the perishing of the process, on the completion of the particular existent, constitutes that existent as an original element in the constitutions of other particular existents elicited by repetitions of process. This kind I have called 'transition.' Concrescence moves towards its final cause, which is its subjective aim; transition is the vehicle of the efficient cause, which is the immortal [i.e., objectively immortal] past" [*PR* 320].

This passage needs to be juxtaposed with another before being interpreted. "There are two species of process, macroscopic process, and microscopic process. The macroscopic process is the transition from attained actuality to actuality in attainment; while the microscopic process is the conversion of conditions which are merely real into determinate actuality [i.e., it is concrescence]" [*PR* 326]. This second quotation makes it clear that these two kinds of fluency are species of process, and it must be emphasized that they are species of *one* process. There are not two processes in Whitehead's system—there is one process, but it is possible to discuss it from two different perspectives, in two different contexts.

To understand how the two kinds of fluency are species of one process, one must understand the principle of creativity, and in what follows the account provided in the CREATIVITY entry of this glossary is presupposed. Process is the creative thrust from *many* to *one,* producing a novel entity that is other than the *many* that gave rise to it and thus part of a new *many* in turn productive of new novel entities. This rhythmic alteration between *many* and *one* is process. Sometimes Whitehead focuses upon the *emergence* of the new *one* that arises from the *many,* and in this case he discusses concrescence, microscopic process, the series of phases in which the emerging actual entity meets what is datum for it with its own creative, telic urge to work through to its satisfaction. At other times he wishes to emphasize the creative advance from faded actual entities to fresh actual entities, the ongoing expansion of the universe, and in this case he discusses transition, macroscopic process, the efficient causation whereby the past affects the future.

"The notion of 'organism' is combined with that of 'process' in a twofold manner. The community of actual things is an organism; but it is not a static organism. It is an incompletion in process of production. Thus the expansion of the universe in respect to actual things is the first meaning of 'process'; and the universe in any stage of its expansion is the first meaning of 'organism.' In this sense, an organism is a nexus. Secondly, each actual entity is itself only describable as an organic process. It repeats in microcosm what the universe is in macrocosm. It is a process proceeding from phase to phase, each phase being the real basis from which its successor proceeds towards the completion of the thing in question" [*PR* 327].

Proposition—A proposition is a hybrid sort of entity in which an eternal object, simple or complex, is fused with an actual entity, or nexus of actual entities. In the fusion both the eternal object and the actual entities involved lose certain of their characteristics. The eternal object, qua eternal object, refers only to the purely general *any* among undetermined actual entities; but as ingredient in a proposition, the eternal object (now called the *predicative pattern* of the proposition) is restricted to just those actual entities with which it is fused, losing its absolute generality of reference, though it still retains its character as merely a potential determinate of actual entities. There is also an abstraction from the completely determinate character of the actual entities involved in the fusion. The concrete definiteness of the actualities is eliminated from—they retain the indicative function of pointing out a particular location, but the eternal objects that in fact characterized them are eliminated so that, in the proposition, they are reduced to bare *its,* bare possibilities for accepting any assigned predicative pattern. As such bare *its* they are referred to as the *logical subjects* of the proposition.

The proposition itself is indeterminate as to its own truth; though because the logical subjects are, as fact, completely determined, the proposition is true, or false, depending upon whether the predicative pattern is, or is not, the actual form of definiteness exemplified in the actual world by the logical subjects. A proposition is also in itself indeterminate as to its own realization in propositional feelings. It is a datum for feeling that awaits a subject to feel it, and its function in the world is to act as a lure for feeling. Whitehead insists over and over that this function is far more important than the simple truth value of a proposition, which has been emphasized by logicians. The ontological principle demands that every entity be somewhere (where *somewhere* means some actual entity) and every proposition has a *locus*. The locus of a proposition is constituted by all those actual entities whose actual worlds include the logical subjects of the proposition, though not all the actual entities in the locus of a proposition will prehend the proposition positively. Many people in a given town may be aware of the existence of an empty lot in the center of town, but only one enterprising businessman may positively prehend the proposition indicated by the words *restaurant on that corner*. At the moment he first prehends the proposition, it is false. But this is not the important fact about the proposition. As a lure for feeling the proposition may lead the businessman to buy the lot and build the restaurant. This is the important function of propositions; they pave the way for the advance into novelty.

Propositional feeling—A propositional feeling is a feeling that has a proposition as its datum. A propositional feeling is a simple comparative feeling (*see Figure 2 of this book*). "An eternal object realized in respect to its pure potentiality as related to *determinate* logical subjects is termed a 'propositional feeling' in the mentality of the actual occasion in question" [*PR* 326]. (*See also* PROPOSITION *and* PHYSICAL PURPOSE.)

Regnant society—*See* Structured society.

Reversion—*See* Conceptual reversion.

Satisfaction—"The actual entity terminates its becoming in one complex feeling involving a completely determinate bond with every item in the universe, the bond being either a positive or a negative prehension. This termination is the 'satisfaction' of the actual entity" [*PR* 71]. *Satisfaction* embodies "the notion of the 'entity as concrete' abstracted from the 'process of concrescence'; it is the outcome separated from the process, thereby losing the actuality of the atomic entity, which is both process and outcome. . . . The 'satisfaction' is the 'superject' rather than the 'substance' or the 'subject.' It closes up the entity; and yet is the superject adding its character to the creativity whereby there is a becoming of entities superseding the one in question" [*PR* 129]. This closing up of the actual entity "embodies what the actual entity is beyond itself" [*PR* 335]—i.e., it is that factor whereby an actual entity "adds a determinate condition to the settlement for the future beyond itself" [*PR* 227]. (*See also* OBJECTIVÉ, SUPERJECT, *and* OBJECTIVE IMMORTALITY.)

Simple feeling—*See* Primary feeling.

Society—A society is a nexus with social order. (*See* NEXUS *for a general account.*) "A nexus enjoys 'social order' where (i) there is a common element of form illustrated in the definiteness of each of its included entities, and (ii) this common element of form arises in each member of the nexus by reason of the conditions imposed upon it by its prehensions of some other members of the nexus, and (iii) these prehensions impose that condition of reproduction by reason of their inclusion of positive feelings of that common form" [*PR* 50–51]. This common element of form is called the *defining characteristic* of the society.

The import of the formal definition emerges clearly in the following passage: "The point of a 'society,' as the term is here used, is that it is self-sustaining; in other words it is its own reason. Thus a society is more than a set of entities to which the same class-name applies: that is to say, it involves more than a merely mathematical conception of 'order.' To constitute a society, the class-name has got to apply to each member, by reason of genetic derivation from other members of that same society. The members of the society are alike because, by reason of their common character, they impose on other members of the society the conditions which lead to that likeness" [*PR* 137].

There are many types of societies and grades of complexity of societies, many societies of different sorts within societies of societies. (*For further information see* CORPUSCULAR SOCIETY, ENDURING OBJECT, *and* STRUCTURED SOCIETY.)

Structured society—A structured society is a complex society that includes subordinate societies and/or nexūs. "A structured society consists in the patterned intertwining of various nexūs with markedly diverse defining characteristics. Some of these nexūs are of lower types than others, and some will be of markedly higher types. There will be the 'subservient' nexūs and the 'regnant' nexūs within the same structured society. This structured society will provide the immediate environment which sustains each of its sub-societies, subservient and regnant alike" [*PR* 157].

Some structured societies are termed *inorganic*. Such societies are complex, indeed, but they have not reached the full heights of complexity. At this level occasions within the society, as they prehend the world about them, elicit a "massive average objectification of a nexus, while eliminating the detailed diversities of the various members of the nexus in question" [*PR* 154]. Here we are at the level of "crystals, rocks, planets, and suns" [*PR* 155], and we are dealing with "the intervention of mentality operating in accordance with the Category of Transmutation" [*PR* 154].

A living structured society is more complex. It will have inorganic structured societies internal to itself as component nexūs, but these will be subservient nexūs. The regnant nexūs in a living structured society will be living. A nexus is living when it includes some living occasions; "thus a society may be more or less 'living' according to the prevalence in it of living occasions" [*PR* 156]. A living occasion is an actual entity that generates "initiative in conceptual prehensions, i.e., in appetition" [*PR* 155]. Such occasions, strung into a society, do not simply elicit a mass average and ignore details; they originate novelty to match the novelty of the environment. "In the case of the higher organisms, this conceptual initiative amounts to *thinking* about the diverse experiences; in the case of lower organisms this conceptual initiative merely amounts to thoughtless adjust-

ment of aesthetic emphasis in obedience to an ideal of harmony"
[*PR* 155]. "In accordance with this doctrine of 'life,' the primary
meaning of 'life' is the origination of conceptual novelty—novelty
of appetition" [*PR* 156].

Tomatoes and individual cells are living; the full account of
the living animal or living human being requires yet further
complications. "In a living society only some of its nexūs will be
such that the mental poles of all their members have any *original*
reactions. These will be its 'entirely living' nexūs, and in practice
a society is only called 'living' when such nexūs are regnant"
[*PR* 157]. The entirely living nexus is a subordinate nexus, not a
subordinate society—i.e., it requires the protection of the whole
society if it is to survive. The reason for this is crucial. The
entirely living nexus is *not* a corpuscular society—i.e., it is *not*
composed of strands of enduring objects. Enduring objects are
personally ordered; their past bears in on them, whereas life
must be free of such confinements—"Life is a bid for freedom:
an enduring entity binds any one of its occasions to the line of its
ancestry. The doctrine of the enduring soul with its permanent
characteristics is exactly the irrelevant answer to the problem
which life presents" [*PR* 159]. The entirely living nexus enjoys
intense experience "derivate from the complex order of the
material animal body, and not from the simple 'personal order'
of past occasions with analogous experience. There is intense
experience without the shackle of reiteration from the past. This
is the condition for spontaneity of conceptual reaction" [*PR* 161].
In summary, an entirely living nexus inherits primarily from the
complex environment provided by the animal body, and not from
its own previous generations. But an entirely living nexus, though
nonsocial in this sense, "may support a thread of personal order
along some historical route of its members. Such an enduring
entity is a 'living person.' It is not of the essence of life to be a
living person. Indeed a living person requires that its immediate
environment be a living, non-social nexus" [*PR* 163]. This notion
of a living person is what answers to the traditional concept of a
soul; the presupposed nonsocial nexus wanders from part to part
of the brain, inheriting from those parts of the brain reporting
bodily activity, whereas the living person is the locus of unified

central control that is regnant over the entire complex structured society that is the living creature.

Subjective aim—The subjective aim of an actual entity is the ideal of what that subject could become, which shapes the very nature of the becoming subject. The doctrine that each actual entity is *causa sui* means that there is not first a subject, which then sorts out feelings; it means, rather, that there are first feelings, which, through integrations, acquire the unity of a subject. Process doesn't presuppose a subject; rather, the subject emerges from the process. "This doctrine of the inherence of the subject in the process of its production requires that in the primary phase of the subjective process there be a conceptual feeling of subjective aim: the physical and other feelings originate as steps towards realizing this conceptual aim through their treatment of initial data" [*PR* 342].

This subjective aim arises in the primary phase of each actual entity as a result of its hybrid physical feeling of God. As primordial, God prehends conceptually the realm of eternal objects; as consequent, he prehends physically the actualities of the world as they arise. Through integrations of feeling God weaves his physical feelings upon his primordial vision, bringing into prominence relevant novel possibilities for the world at its given stage of attainment. As superject, God offers for each actual entity, as its subjective aim, a vision of what that entity might become. This subjective aim constitutes the ideal for growth on the part of each actual entity that would result in maximum ordered complexity in the world were it realized in fact—this is God's mode of operation in the world, designed to produce the kind of world that, physically prehended by his consequent nature, would result in maximum intensity of satisfaction for him. Subjective aims, then, constitute the means by which God works in the world. When Whitehead states that he does not conceive of God under the image of the ruling Caesar, or the ruthless moralist, or the unmoved mover, but rather in the spirit of the "brief Galilean vision of humility" that "dwells upon the tender elements in the world, which slowly and in quietness operate by love" [*PR* 520], he is implicitly referring to his doctrine of God as the source of subjective aims. God works slowly because there is no compulsion upon an actual entity to accept the proffered lure—it is possible for a subjective aim to suffer simplification and modification in the successive phases of concrescence. (*See also* GOD.)

Subjective form—The subjective form of a prehension is *how* the subject of the prehension feels the datum of that prehension. (*See* PREHENSION.) A subjective form is the subjective form it is because it has a specific definiteness constituted by the eternal objects (of the subjective species) that are ingredient in it. "There are many species of subjective forms, such as emotions, valuations, purposes, adversions, aversions, consciousness, etc." [*PR* 35]. The same datum may be received in different subjects clothed in very different subjective forms; as, for example, a permissive mother and an elderly spinster neighbor may view the selfsame shenanigans of energetic children in quite a different light, with quite a different emotional response. Valuations are the subjective forms of conceptual feelings, and are either valuation up (adversion) or valuation down (aversion): in the former case they insure the continued importance of their data in the present concrescence and beyond in the future; in the latter case there is some degree of attenuation of importance of the data.

The subjective forms of the prehensions constituting a given subject do not arise independently. They influence each other, and their over-all character is determined by the one subjective aim (*which see*) that dominates the self-formation of the concrescing subject.

Superject—"An actual entity is to be conceived both as a subject presiding over its own immediacy of becoming, and a superject which is the atomic creature exercising its function of objective immortality. . . . It is subject-superject, and neither half of this description can for a moment be lost sight of" [*PR* 71, 43]. The superjective character of an actual entity "is the pragmatic value of its specific satisfaction qualifying the transcendent creativity" [*PR* 134]—i.e., it is that character it has as dead datum functioning as a given object for the concrescence of subsequent generations of actual entities. (*See also* SATISFACTION, OBJECTIVÉ, *and* OBJECTIVE IMMORTALITY.)

Symbolic reference—Symbolic reference is the mixed mode of perception characteristic of fully alert human perception. It is an integration of, an interplay between, perception in the mode of causal efficacy and perception in the mode of presentational immediacy.

Perception in the mode of causal efficacy "is perception of the settled world in the past as constituted by its feeling-tones, and as efficacious by reason of those feeling-tones" [*PR* 184]. This mode of feeling permeates the physical world and is exemplified in human experience by visceral feelings—a nagging stomach ache, for instance—or by the brute givenness of memory. Perception in the mode of presentational immediacy, on the other hand, is "perception which merely, by means of a sensum, rescues from vagueness a contemporary spatial region, in respect to its spatial shape and its spatial perspective from the percipient . . ." [*PR* 185]. Presentational immediacy is the product of *bare sight* —"But we all know that the mere sight involved, in the perception of the grey stone, is the sight of a grey shape contemporaneous with the percipient, and with certain spatial relations to the percipient, more or less vaguely defined. Thus the mere sight is confined to the illustration of the geometrical perspective relatedness, of a certain contemporary spatial region, to the percipient, the illustration being effected by the mediation of 'grey.' The sensum 'grey' rescues that region from its vague confusion with other regions" [*PR* 185].

Whitehead argues that philosophers, in analyzing perception, have tended to ignore perception in the mode of causal efficacy. "Philosophers have disdained the information about the universe obtained through their visceral feelings, and have concentrated on visual feelings" [*PR* 184]. The result is that philosophy has attempted to analyze perception in terms solely of presentational immediacy, and this has led to Hume's skepticism. "Hume's polemic respecting causation is, in fact, one prolonged, convincing argument that pure presentational immediacy does not disclose any causal influence. . . .The conclusion is that, in so far as concerns their disclosure by presentational immediacy, actual entities in the contemporary universe are causally independent of each other ' [*PR* 188].

But, Whitehead holds, there *is* a *causal influence* that permeates ordinary perception. "When we register in consciousness our visual perception of a grey stone, something more than bare sight is meant. The 'stone' has a reference to its past. A 'stone' has certainly a history, and probably a future, when it could be used as a missile if small enough, or as a seat if large enough" [*PR* 184–185]. The sensa involved in presentational immediacy have been derived from primitive feelings in the prior animal body, from perception in the mode of causal efficacy; and the vague massiveness of the presence of the past that originally accompanied them in that mode is not totally lost when they are projected onto a sharply defined contemporary spatial region in the mode of presentational immediacy, so that the mixed mode of symbolic reference perceives the stone both as clearly located in a contemporary region of space and yet as also a persisting entity with a past and an efficacy in the future. (*See also* CAUSAL EFFICACY, PRESENTATIONAL IMMEDIACY, *and* CONSCIOUSNESS.)

Theory—*See* Proposition.

Transmutation—Transmutation is the operation whereby macrocosmic perceptions arise out of microcosmic prehensions—e.g., whereby perception of *a* table supercedes the prehension of the welter of individual actual entities constitutive of the table. Transmutation is possible when in a series of pure physical feelings in a given subject (these feelings being feelings of the individual members of a nexus of actual entities exhibiting some important characteristic of order) there is identity of pattern in their ingredient eternal objects. In the second phase of that given subject the same identical conceptual feeling arises from each of these physical feelings. This one conceptual feeling, heavily reinforced and impartially relevant to all, or the vast majority of, the pure physical feelings of the first phase, is the key to transmutation. In the integrated feeling that follows in Phase III, this one conceptual feeling is set in contrast to the nexus as a whole, and "the nexus as a whole derives a character which in some way belongs to its various members" [*PR* 386]. This transmuted physical feeling of the nexus as a whole exemplifying some given characteristic has substituted one feeling of the nexus in place of the initially many feelings of the various actual entities of the nexus—there is the feeling of a table as over against many feelings of the actual entities constitutive of the table. (*See Figure 3 of this book and accompanying text.*)

Ultimate, the—*See* Creativity.

Valuation—*See* Conceptual prehension *and* Subjective form.

Source Key

LOCATION IN *PROCESS AND REALITY* OF PASSAGES QUOTED IN THIS BOOK

The first, boldface number in each line indicates a page in this book, and the subsequent number or numbers in regular type indicate the source in *Process and Reality* of all material from *PR* found on that page. For example, the line

16   234, 339, viii–ix

means that page 16 of this book is quoting from page 234 of *PR* when it commences at the top, that when it ceases to quote page 234 it has moved to a passage from *PR* 339, and that at the bottom it is quoting a passage from *PR* which spans pages viii and ix. Slashes have been used to indicate the position on pages in this book, relative to the quoted passages, of any break in the text, such as a section head or an italicized insertion by the editor. Thus the line

17   ix, 94, 129, 129/ 36, 37

means that the material on page 17 of this book *above* the section head is taken, in sequence, from pages ix, 94, and 129 of *PR,* while the material *below* the section head is taken, in sequence, from pages 36 and 37 of *PR.* The double occurrence of 129 means that two non-sequential passages from *PR* 129 occur together in this book. The boldface numbers indicating page numbers of this book are in sequence, but since some pages are completely given over to explanatory material and do not quote from *PR,* their numbers are missing from the sequence.

CHAPTER 1

7 viii, 27–28, 168, 28, 116, 321, 323, 335

8 124, 65, 117–118, 290, 65, 249, 420, 133, 322, 334, 335, 81, 65, 66

9 66, 65–66, 66, 338, 339/ 35/ 361

10 361

11 361, 470, 97, 353, 338, 338, 353–354, 361, 91, 363, 363

12 364/ 337–338, 234, 35, 354, 356, 356, 354

13 479–480, 362, 357, 362–363, 355, 363, 363

14 335, 71, 355, 322, 322, 340, 352, 336, 336, 129

15 129, 336, 336, 335, 336, 336, 336, 71, 89, 43, 339, 234

16 234, 339, viii–ix

17 ix, 94, 129, 129/ 36, 37

18 28, 113, 254, 321, 116

19 116, 68–69

CHAPTER 2

20 63

21 63, 70, 226, 70, 70, 72, 38, 34

22 34, 44, 226, 35, 367, 366–367

23 367, 366/ 33

24 33, 43, 79–80, 38, 80, 230, 225, 93, 78, 364

25 364, 249/ 73, 73, 73, 73, 63–64

26 64, 64, 392, 392, 392, 48, 378, 48, 46, 522

27 522, 248, 522, 64, 46, 46, 46, 47

28 47/ 342, 373, 373, 343, 373, 374

29 374, 374/ 521, 168

30 168, 134, 134, 134, 134, 134, 134, 134, 134–135, 135

31 135/ 104, 522, 287, 522, 377, 135, 161, 373–374, 343–344

32 344, 521, 344/ 10–11, 339, 11, 46–47

33 47, 339, 339–340/ 31, 31, 31, 31, 31–32, 89

34 89, 31, 32, 321–322, 327, 322, 32, 32, 438, 32, 443, 32, 229, 348

CHAPTER 3

36 337, 433–434, 434, 434, 434

37 347, 105–106, 359–360

38 360, 335

39 39, 39, 249, 249, 251

40 230, 362, 375, 323, 101.

41 101, 363, 363–364, 362, 364, 250

42 340, 341, 364, 362, 347, 364

43 364, 338, 344/ 345, 345

44 345–346

45 346, 347, 348, 347

46 366, 165, 366, 165, 378

47 379, 379/ 380

48 380, 40, 380–381, 381–382

49 375, 375, 376, 377, 377, 343

50 343, 377, 382, 377

51 367, 388, 377, 368, 369, 368

52 389, 41, 389, 41, 381, 389–390, 390, 390

53 390, 424, 41, 424, 390, 390, 424

54 424–425

55 365–366, 380, 380, 421

56 421–422, 422–423, 388

57 388, 423, 374–375

58 375, 375–376, 269, 375, 423–424, 425–426

59 426, 285, 426/ 391, 391, 391

60 391–392, 392–393

61 393–394, 398, 394, 398–399

62 394–395, 326, 286, 395, 286

63 283, 284, 283–284

64 284, 281, 395, 281, 395–396, 281, 37

65 37, 37, 395

66 280, 421, 280, 402, 402

67 427–428/ 407, 407, 407, 407, 408

Index